Product Customization

Lars Hvam · Niels Henrik Mortensen · Jesper Riis

Product Customization

 Springer

Lars Hvam
Technical University of Denmark
Building 425
DK2800 Lyngby
Denmark

Jesper Riis
GEA Niro
Gladsaxevej 305
DK2860 Søborg
Denmark

Niels Henrik Mortensen
Technical University of Denmark
Building 404
DK2800 Lyngby
Denmark

ISBN 978-3-642-09064-6   e-ISBN 978-3-540-71449-1

DOI 10.1007/978-3-540-71449-1

Cover design: WMX Design GmbH, Heidelberg

Printed on acid-free paper

9 8 7 6 5 4 3 2 1

springer.com

Preface

This book describes a procedure for designing configuration systems. These systems are used to an increasing extent in industry to support the task of specifying products seen in relation to, for example, sales and production. The procedure involves analysis and redesign of the business processes which can be supported by a configuration system, analysis and modelling of the company's product range, selection of configuration software, programming the software, and implementation and further development of the configuration system.

In this book we have chosen to focus most of our attention on the initial phases of the procedure, involving analysis and redesign of business processes and modelling of the product range. The book covers a relatively large number of technical disciplines, which we have attempted to deal with by dividing up the content among us. The authors of the individual chapters are: 1, 2, 4 and 9: Lars Hvam, 5: Niels Henrik Mortensen and Lars Hvam, 3: Lars Hvam, Jesper Riis and Niels Henrik Mortensen, 6, 7 and 8: Jesper Riis and Lars Hvam.

Chapter 9 describes some experiences with the use of configuration systems at F.L. Smidth A/S. In this connection we would like to thank Line Hemmingsen, Michael K. Nielsen and Morten Hugo Bennick for an excellent collaboration, and for setting aside the time to discuss their experiences with product configuration and sharing them with us.

The starting point for the procedure which we describe here was Lars Hvam's Ph.D. project, "Application of product modelling – seen from a work preparation viewpoint". The original procedure has subsequently been developed further in Ph.D. projects carried out by Benjamin Loer Hansen, Jesper Riis and Martin Malis, under the supervision of Lars Hvam, by Carsten Svensson under the supervision of Ari Barfod and Lars Hvam, and by Ulf Harlou under the supervision of Niels Henrik Mortensen.

A vital factor in the development of the procedure described here has been that we have had a close collaboration over the last 10 years with a number industry and service companies on the design and use of configuration systems. This collaboration has been mediated through the Ph.D.

projects mentioned above, and by more than 60 M.Sc. thesis projects supervised by Niels Henrik Mortensen and Lars Hvam.

We would like to thank Robin Sharp of Translator Data for translating
the book into English and Morten Kvist for producing the final version of
the figures and the layout.

Lyngby, August 2007

Lars Hvam   Niels Henrik Mortensen   Jesper Riis

Table of Contents

Abbreviations

AI	Artificial Intelligence	
API	Application Programming Interface	Interface for communication with external IT systems
AS-IS model		Model of an existing system
BPR	Business Process Reengineering	Method for development of business processes
CAD	Computer Aided Design	IT systems for drawing products in 2D or 3D
CASE	Computer Aided Software Engineering	Systems for supporting development and maintenance of IT systems
CIM-OSA	Computer Integrated Manufacturing Open System Architecture	Reference model for software for industrial enterprises
CNC	Computer Numeric Control	Computer-controlled production machines
CRC card	Class Responsibility Collaboration Card	Card for describing object classes
CRM	Customer Relation Management	IT systems for managing customer data
DDE	Dynamic Data Exchange	Communication between different IT systems under Windows Communication between different IT systems under Windows
ERP	Enterprise Resource Planning	IT systems for managing resources – such as materials, machines and personnel – within companies

OCL	Object Constraint Language	Notation for modelling of constraints
ODBC	Open Database Connection	Definition of interface to databases
OLE	Object Linkage & Embedding	Method for importing objects into IT systems under MS-Windows
OOA	Object-oriented analysis	Method for analysis of the application area for an object-oriented IT system
OOD	Object-oriented design	Method for design of object-oriented IT systems
PDM	Product Data Management systems	IT systems for managing product data
PEST	Political, Economical, Sociological, Technological	Description of a company's political, economical, sociological and technological status
SCM	Supply Chain Management	Management of supply chains (suppliers)
SWOT	Strength Weaknesses Opportunities, Threats	Description of a company's strengths, weaknesses, opportunities and threats
TO-BE model		Model of a future system
UI	User Interface	
UML	Unified Modelling Language	Standard technique for modelling object-oriented IT systems

1

Introduction

Many companies are currently experiencing increasing demands from their customers for the delivery of customized products that have almost the same delivery time, price and quality as mass-produced products. One way this development is described is by the concept of mass customization – a production form in which customized products are delivered which exploit the advantages of mass production.

Using the principles of mass customization implies a radical revision of the company's overall business model. Some of the central elements in a mass customization strategy are:

- A focussed market strategy, i.e. a clear strategy for which customers should be serviced with which products, and the will to refuse customers lying outside this segment.
- Sales and development of manufacturing specifications for customized products that utilize product configuration systems.
- A product range based on modules, so a customized product is put together by selecting, combining and possibly adapting a set of standard modules.
- Mass production of standard modules and customer-initiated assembly of customized products based on the use of modules.
- Installation and after-sales service of products based on installation and replacement of modules.

Of course, many different examples exist of how companies in practice develop their business models based on the principles of mass customization. The central element in such a business model, and the element that is new for most companies, is the development of a module-based product range and the use of product configuration systems in customer-oriented business processes that include sales, product design, and development of manufacturing specifications for customized products.

This book focuses on how customer-oriented business processes can be developed through the use of configuration systems. Information and communication technology (ICT) offers a number of new possibilities for developing these business processes, which involve specification of products in relation to customer requirements, development of the manufacturing specifications etc. – in other words, those business processes which lie between the customer and production.

We are currently seeing the first examples of companies that have developed IT systems (so-called configuration systems) to support the development of specifications for customized products. The commercial advantages of doing this can be considerable.

The company American Power Conversion (APC), for example, has in recent years achieved impressive results by using the principles of mass customization and product configuration. This company, which produces infrastructure systems for computer centres, has developed and implemented a coherent business model, which includes a focussed market strategy, development of a module-based product range, sales and order processing based on the use of configuration systems, a production system with customer-initiated assembly of customized products based on standard modules, and distribution, installation and after-sales service based on the use of standard modules in the product range.

A central element in APC's business model is the use of configuration systems which contain a description of the individual modules together with rules for the selection, dimensioning and combination of modules.

APC currently has a graphic configuration framework, which supports 11-12 types of products and therefore covers all the equipment to be used in the infrastructure system for a large computer centre – racks, emergency power supplies, controls, air conditioning, cabling systems etc.

The configuration systems, which are used by more than 10,000 sales staff and agents, make it possible to configure the infrastructure system for a large computer centre and to work out an offer and manufacturing specifications in less than an hour. This is a process which previously took from 3-4 days to 3-4 weeks. At the same time, the overall delivery time has been reduced from 400 days to 16 days. The target is to reduce the delivery time further to 4 days. Moreover, product quality and productivity have been noticeably improved.

Another example is the Swedish company Sandviken, which amongst other things manufacture steel holders. Sandviken has developed configuration systems to support the calculation of price and work out product specifications and manufacturing specifications for customized steel holders.

Previously, Sandviken's staff used 2-3 weeks for working out the offer and the manufacturing specifications for a steel holder. Today, these specifications are produced automatically: A salesman types in a number of parameters describing the customer's requirements for a steel holder, and then the configuration system, which contains rules for working out drawings, lists of parts, CNC codes etc., creates the drawing, parts list and the remainder of the manufacturing specifications (in the form of the list of operations, CNC codes and operation descriptions). In addition, the system calculates the price of the steel holder and prints out an offer. Thus, the time for making the offer and the manufacturing specifications is reduced from 2-3 weeks to 10 minutes. At the same time, the quality of the specifications is increased.

A third example is F.L. Smidth, which manufactures cement factories. F.L. Smidth currently has more than 30% of the world market for equipment for cement manufacture. F.L. Smidth used many resources to work out offers for their customers. The company has just implemented a configuration system to support the calculation of budgetary offers. Until now, the time required for making a budgetary offer has been 3 to 5 weeks. In addition to the salesman, 10-15 specialists, who have contributed knowledge about various parts of the cement factory, have typically been involved. This knowledge has now been stored in the configuration system. This means that a salesman can work out a budgetary offer for a customer on his own in the course of 1 to 2 days by using the configuration system.

The system inspires a vastly improved dialogue with the customers, since the short response time makes it possible to simulate different solutions, according to the customer's requirements. In addition, it becomes possible to optimize the plant with respect to F.L. Smidth's preferences for using standard modules, choice of suppliers etc. This leads to a reduction in F.L. Smidth's costs for subsequent detailed engineering, production, and setting up of the cement factory.

As the examples demonstrate, the use of configuration systems can lead to marked improvements in the customer-oriented business processes related to the specification of customized products. Some of the effects which can typically be achieved are as follows:

- Reduced time for working out specifications, for example from weeks to hours or minutes.
- Faster and more qualified responses to customer enquiries. For example, it becomes much faster and requires significantly fewer resources to give a customer an offer.
- Fewer transfers of responsibility and fewer errors in the specifications worked out.
- Reduced use of resources for specifying customized products, and therefore more time, for example, for product development and sales work.
- Possibility of optimizing products in relation to the customer's needs, and for reducing costs, for example for materials and production.

It is a question of radically reorganising the business processes that form the connection between the customer and the production system. A common feature of the examples we see today is that knowledge and information about products and their production, assembly etc. is formalized and put into an IT system (the configuration system). Thus, a business process is created which has a high degree of formalization, a high degree of IT support, and markedly improved performance with respect to predictability, efficiency and quality. To illustrate the possibilities in this, let us consider an imaginary example from an industrial company.

Doors Inc.

Doors Inc. manufactures doors for agricultural and industrial buildings. The doors are adapted to the individual customer with respect to dimensions, degree of insulation, choice of materials, colour etc. For a number of years, the company has worked on rationalizing and improving their production. One result is that production time for standard doors has been reduced to 2 weeks, while productivity and quality have both improved considerably.

The customers, however, have not experienced any great improvement. Delivery time for standard doors is currently 6-8 weeks. It takes 3-5 weeks from the time the customer has given an order till the finished manufacturing specifications in the form of drawings, lists of parts, lists of operations etc. are available. The final week is used for purchasing materials and planning and printing out production instructions. Out of an overall workforce of rather more than 250 employees, about 10 people are employed in dealing with orders and working out the manufacturing specifications. On top of this comes the time used by the salesmen and other employees in the production engineering and product development departments for specification of customer orders.

The fitters who install the doors on the customers' premises have calculated that in more than 20% of the orders, the dimensions of the gate are incorrect or some parts of the gate are missing. In both cases, considerable delays and extra costs are involved. At the same time, the employees in the production department complain that the specifications on which production relies, such as drawings, lists of parts, operation lists etc., contain far too many errors. A few samples have indicated that more than 30% of the specifications used contain errors.

Doors Inc. faces increasing market demands for rapid delivery, at the same time as it is becoming more and more difficult to match their competitors' prices. Over the last couple of years, the company has experienced dwindling sales and reduced profits. Therefore, the company's top management has launched an initiative aimed at improving the way in which the company deals with customer orders. The target is to reduce the delivery time for standard doors to 3 weeks. The company will try to achieve this goal by reducing the time from receipt of order to start of production to one week. At the same time, the cost price is to be reduced by 10% a year over the next 2 years. It is estimated that the main part of the price reduction target can be achieved by reducing the number of faulty specifications. A further target is to increase the number of offers by 20% a year with an unchanged hit rate.

An analysis has been carried out of current business processes related to the production of offers and the handling of customer orders. Figure 1.1 shows the sequence of events from the sales contact and negotiations with the customer until the gate has been installed and invoiced.

As shown in figure 1.1, many different people are involved in selling, designing, producing and installing a gate for a customer. Before the production department starts to saw the first profile for the gate, 4 to 6 weeks have already passed since the customer gave the order. More than 10 different people have been involved in dealing with the order, the customer's wishes have been reformulated many times, and in all probability, errors have been introduced into some of the specifications sent to production.

Taking the targets and the analysis of the current business process as

Figure 1.1.
A customer order on its way through Doors Inc.

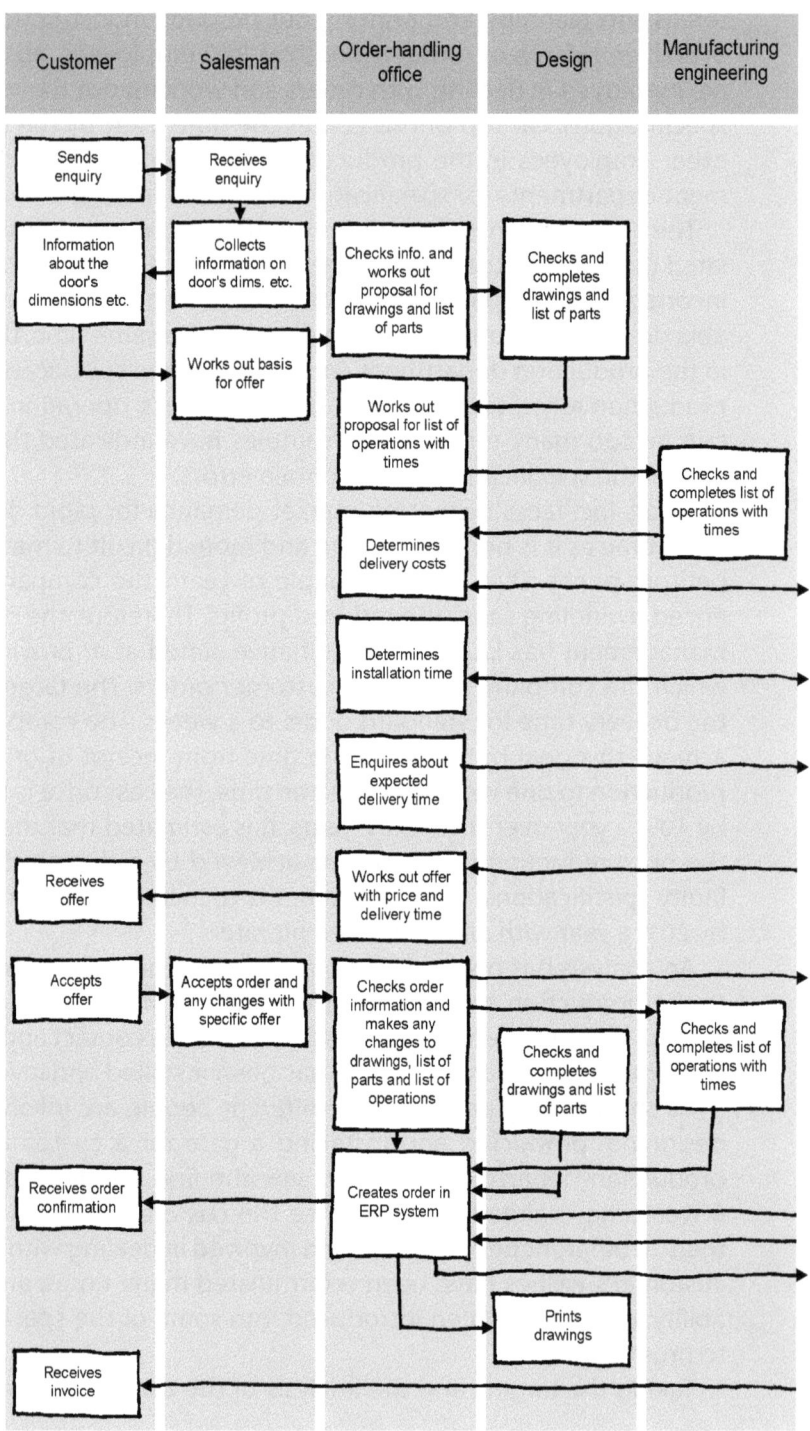

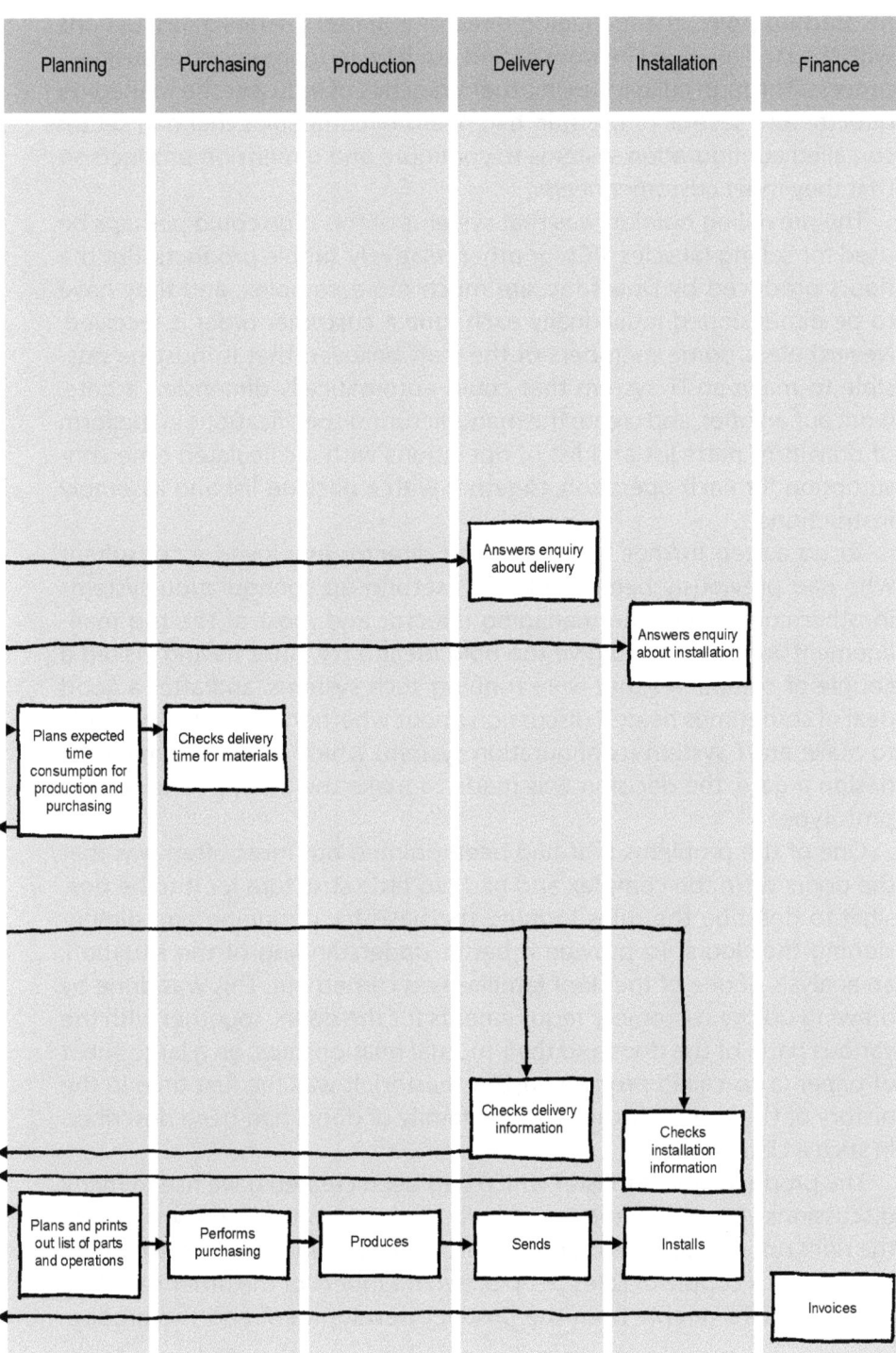

his starting point, the managing director initiated a series of discussions with the staff involved in order to find out how to reorganise the business process. Through colleagues in other branches of industry, the managing director and several of the staff had heard of companies that had set up so-called configuration systems to configure and dimension products so that they meet customer needs.

The prevailing opinion was that systems of this type could perhaps be used for selling bicycles, PCs, or other relatively simple products. But the doors produced by Doors Inc. are much more complex, and they have to be dimensioned individually each time a customer order is received. Nevertheless, some members of the staff believed that it must be possible to make an IT system that could automatically dimension a gate, work out an offer, and create the manufacturing specifications in the form of drawings, parts list and list of operations with a calculated time consumption for each operation, together with a packing list and assembly instructions.

To go a step further, the managing director employed a consultant who had previously been involved in setting up configuration systems in other companies. The managing director and most of the top management were willing to give the bold ideas a try. After having visited a couple of companies that were running such systems, and after a good deal of sometimes heated discussion about whether it was at all possible to make an IT system (configuration system) which could automatically design a gate, the decision was made to make the attempt and build a prototype.

One of the problems that had been pointed out most often was that the doors were too complex and had too little structure for it to be possible to describe the rules forming the basis for designing and dimensioning the doors. To provide a better understanding of the situation, an analysis of one of the door families was carried out. This was done by drawing up the customers' requirements for the doors, together with the various parts of the door and their mutual relationships, on a large sheet of paper (a so-called product variant master). It was the first time in the history of the company that a whole family of doors had been described in such a clear manner.

The product variant master which had been created gave rise to many discussions about what the door family really comprised, and which were the right rules for dimensioning a door. After 4 or 5 meetings between the consultant, a couple of sales people, two members of the order handling team, and two people from the product development and the production engineering departments, respectively, a description was formulated which everyone agreed on. Everyone also agreed that it had been a pleasant experience for the very first time to have a complete overall view of a whole product family.

Doors Inc. had earlier been visited by a company which delivered software for setting up configuration systems. It was agreed to borrow this company's software. One of the staff from the product development department took a 3-day course in programming the system, and then made a prototype of the configuration system. This took 14 days in collaboration with the consultant.

With the prototype, a salesman was able to design a door himself, so it matched the customer's needs. The salesman typed in the dimensions, choice of materials, colour, selected modules, lifting mechanism etc., after which the configuration system automatically generated a list of parts, a list of operations with calculated production times for planning purposes, and a packing list. In addition, the system could print out a letter with the offer to the customer.

Once the prototype was finished, discussions about how to specify doors in the future became much more precise. It was now clear to almost everybody that it really was possible to make an IT system which could automatically specify a door. On the other hand, it was also clear that the system could not possibly handle all conceivable variants of a door. The prototype showed that a realistic target would be for the system to be able to handle 80% of the customer orders within the chosen door family. The remaining 20% would have to be handled separately as special orders.

The sales staff were somewhat positive about the new system. On the one hand, they were enthusiastic that they could now deliver an offer to the customer on the spot, without having to ask product developers and production staff. In addition, a 3-week delivery time represented a breakthrough in the market in relation to the delivery times that were otherwise the norm. On the other hand, many of the salesmen had reservations with respect to whether the system could handle the requirements customers might have.

The staff in the product development and production engineering departments were enthusiastic about the prospect of being able to avoid dealing with the eternal queries from the sales staff. Now, a bit more time could hopefully be available for the development work that had been neglected for so long. At the same time, they were somewhat uncertain with respect to how the knowledge about doors that was incorporated in the configuration system could be kept updated.

The staff in the order handling office were probably the ones who had most reservations. Although there would continue to be a number of special orders that needed dealing with, most of the work involved in specifying the individual customer orders would be handled automatically in the future. The managing director had to admit that the staff in the order handling office would have to be reduced. On the other hand, over the last couple of years 3 or 4 employees from product development and production, who had detailed knowledge of the products and their production, had been moved to the order handling office. There was a considerable need to move these employees back to their original departments, where there was a constant shortage of staff with knowledge of the products and understanding of their production.

The managing director summarized the changes which would take place by saying that they were now going to "industralize" the business processes lying between the customer and production. In other words, the business processes which are related to identification of the customer's needs, design of a door, formulation of the manufacturing specifications, and calculation of the price. In using the word industrialization, he meant that they should prepare this "specification task" by formalizing the knowledge forming the basis for specification of a door, and by exploiting IT to carry out parts of the task.

In addition, he wanted with the use of the term industrialization to express the idea that in the future it would be possible, through the introduction of a high degree of formalization and IT support, to carry out the customer-oriented specification task with greater efficiency, a large degree of predictability, and high and uniform quality. Thus, customers would receive faster and more qualified response to queries and orders. It would also be easier for the salesman, in collaboration with the customer, to design a door corresponding to the customer's needs, since it would be easy to simulate different solutions and then select the optimal design.

On the basis of the experience provided by the prototype, it was decided to continue the work. A project organisation was established and given the task of developing the future business process and the configuration system to be used to support this process. The first step in this task was to define the future business process, which included describing the subtasks making up the process, and to define who had responsibility for the individual subtasks. In this context, an important task was to identify the people who would be responsible for maintenance and further development of the configuration system, which contains knowledge and information about the doors and their manufacturing process. In addition, an overall description was made of the configuration system to be constructed and its integration into the company's ERP system.

The next step was to continue with an analysis of the remaining door families to be incorporated into the configuration system. As in the development of the prototype, the product families were analysed with the help of so-called product variant masters. Details which could not be included in the product variant master were described on a series of associated cards. Developing product variant masters for the various door families gave rise to much discussion about which products and market segments should be included in Doors Inc.'s standard products. After the individual product families had been analysed and described, the next step was to incorporate the product model into the standard software for product configuration that the company had just purchased.

Product analysis, programming of the configuration system, test and debugging took 4 months in total. After this, the way was open to implementation of the system within the organisation. As this was a new technology, which would lead to marked changes in the pattern of work, it was decided to carry out a gradual implementation procedure over a 6-month period. Initially, the two salesmen who had been involved in creating the prototype were trained to use the system. Based on their experiences, a series of adjustments to the system were made over the next few months, and then the rest of the Danish sales force was trained and began to use the system after about 3 months. Finally, the system was implemented in the three foreign subsidiaries.

Figure 1.2 shows the new process for specification and manufacture of a door at Doors Inc. The salesman types in the desired parameters for the door, after which the configuration system designs the door, creates an offer, a list of parts, a list of operations with calculated time consumption for each operation, a packing list and instructions for assembly. The dimensions of the door are transferred to the company's CAD system which produces a drawing of the door. The lists of parts and operations are transferred to the company's ERP system, which then works out a materials and capacity plan for the production and initiates purchasing and production.

With the new business process, only the salesman and the production planner are involved in specification of the customized door, preparation of the offer, and creation of the manufacturing specifications. For standard products, purchasing takes place independently of the individual orders through max./min. stocks of materials and components used in manu-

Figure 1.2.
The new process for producing offers and executing orders for standard doors in Doors Inc.

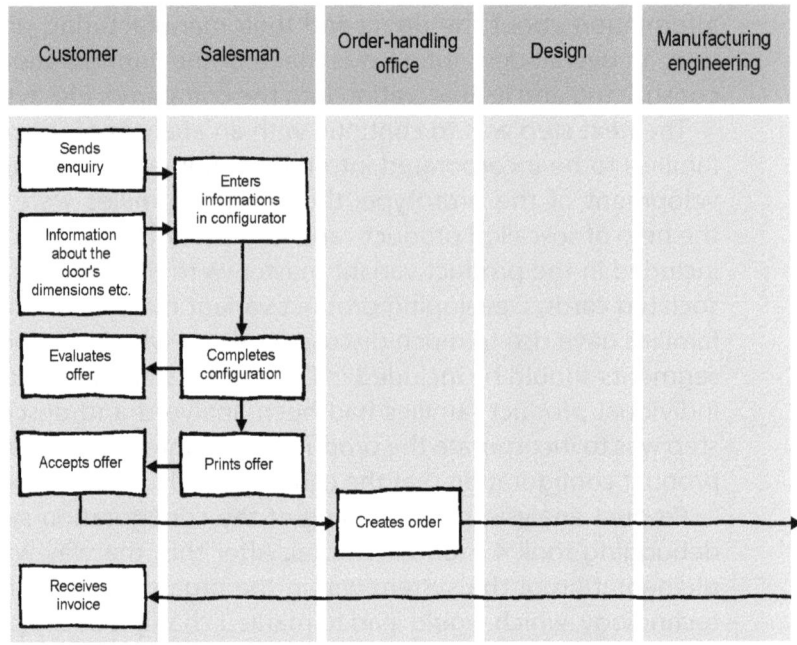

facturing doors within the standard programme. In this way the distance between customer and production has been reduced considerably.

The lead time required for creating specifications has now been reduced to 0.5 - 2 hours, corresponding to the time it takes the salesman and customer together to design the door and create an offer. During the first few months of the system's lifetime, a number of errors were discovered in the configuration system. Once these errors were corrected, the specifications delivered for production, transport and assembly have been almost completely free of errors. This has meant a reduction in costs of at least 20%. In addition, the production department has become much better at keeping to the agreed delivery deadlines.

The staff of the order handling office has been reduced to 3 employees, who now exclusively deal with special orders. The sales force has much more time for actual sales work and has now started to cultivate the Swedish market, which they have talked about doing for several years. The staff in the product development and production engineering departments have more time for development work and have started to develop a completely new family of doors, which are intended to replace 2/3 of the existing product portfolio. In this work, the product variant master is used as early as the development phase, in order to document the new door programme.

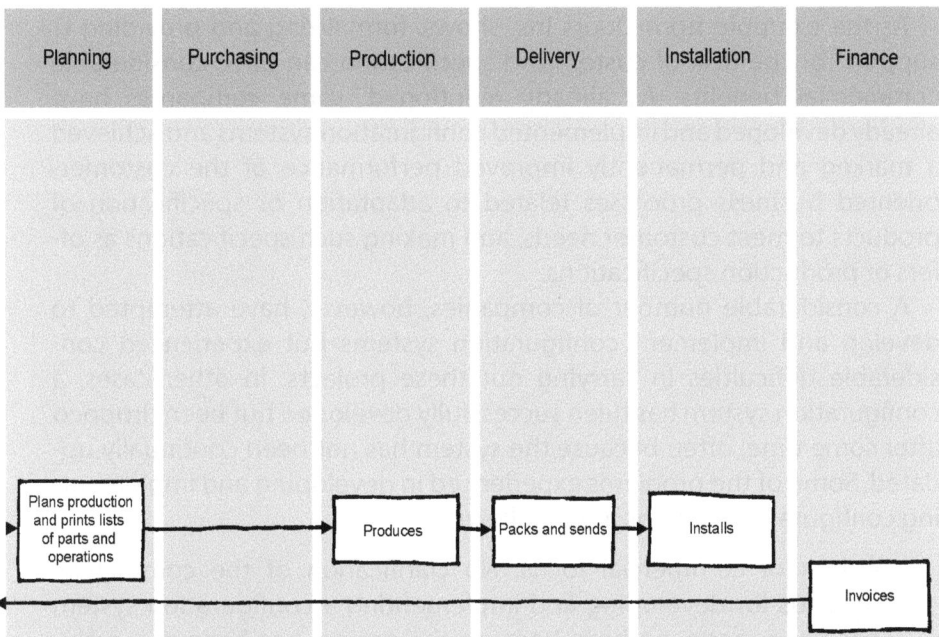

The head of the development department has summarized the new situation by saying that they now have more time for actual development work. On the other hand, the task of development has also become more extensive. It is not only necessary to develop a product concept, as was previously the case, but instead a detailed product programme, in which all details are fully documented. Moreover, the product developmentdepartment has taken over responsibility for maintenance and further development of the configuration system, including building a configuration system for the new door family.

The sales force was initially somewhat worried about whether the configuration system could really deal with the customers' wishes. After using the system for some months, most of the salesmen agreed, however, that the configuration system gave a clear and unambiguous definition of what were standard orders and what were special orders. For special orders, 20% would be added to the price, and the delivery time would be about 10 weeks. This meant that many customers preferred a door which lay within the limits of the product model. The short delivery times and the rapid response to customer enquiries had led to an increase in sales of about 20%.

As the example from Doors Inc. shows, formalizing and providing IT support for the task of customized specification can have considerable commercial benefits. As already mentioned, some companies have already developed and implemented configuration systems and achieved a marked and permanently improved performance of the customer-oriented business processes related to adaptation or specification of products to meet customer needs, and making such specifications as offers or production specifications.

A considerable number of companies, however, have attempted to develop and implement configuration systems but experienced considerable difficulties in carrying out these projects. In other cases, a configuration system has been successfully developed but been dropped after some time, often because the system has not been continually updated. Some of the problems experienced in developing and implementing configuration systems are as follows:

- A lack of commercial focus. No clarification of the commercial grounds for developing and implementing a configuration system has been made, perhaps because the project has been run exclusively by technically oriented people who are primarily interested in the technological possibilities.

- Backing from the management is lacking, and the configuration system is not rooted in the way daily operations are organised. This can mean that the configuration system is not continually updated and that the knowledge incorporated into the configuration system rapidly becomes out of date. As a result, no one is any longer interested in using the system.

- The products are not configurable. The individual product families and their possible variations have not been clearly defined. The product range is unstructured, and no consensus exists about which variants should be offered or which market segments should be serviced.

- The configuration system that has been developed is not structured or documented. This makes it difficult or even impossible to maintain and develop the system further. This situation typically arises when people start to programme without having developed and documented an overall description of the configuration system.

The following chapters, describe a procedure for developing and implementing configuration systems. The first step in the procedure is analysis and redesign of the business processes to be supported by the configuration system, including the future organisation of the work. The next step involves analysing and possibly restructuring products which are to be incorporated into the configuration system. The subsequent steps are detailed modelling, programming, implementation and maintenance. Chapter 9 describes an example of a company that has developed and implemented a configuration system.

Before going on to present the procedure for developing configuration systems, the next chapter gives a more detailed description of the business processes involved in the specification of products and services related to the customer needs - the so-called specification processes. After that follows an introductory description of what a configuration system actually is.

2

Specification Processes and Product Configuration

Two of the central principles of mass customization are that product ranges should be developed on the basis of modules, and that configuration systems should be used to support the tasks involved in the customer-oriented business processes related to the specification of customer-specific products.

This chapter describes the business processes lying between the customer and the production process, and the possibilities offered by product configuration for developing these business processes. Starting with the concept of specifications, we introduce the type of business processes that involve the development of specifications for customer tailored products, and which in this book are denoted specification processes.

Next we briefly discuss the development and use of modules, which are an important means for creating customized products and for obtaining greater efficiency in the company's specification processes, and which make up the foundation on which a configuration system can be built. Finally, we introduce and explain the concepts of a configuration system and a product model, and we present a framework for modelling product families. This framework is used in subsequent chapters to delimit and structure the knowledge which has to be put into a configuration system.

Specifications

Specifications are a concept which we all know from everyday life. A specification can be defined as a description which can unambiguously transfer needs or intentions from one group of people to another. Examples of specifications include baking recipes, assembly instructions for an item of furniture from IKEA, or directions for driving somewhere.

In industrial companies where many people are involved in developing, marketing, selling, producing and servicing products, specifications make up an important part of daily life. Descriptions of customer requirements, product drawings, lists of parts, assembly instructions and service manuals are examples of specifications in industrial companies.

In this book we are particularly interested in those specifications which are created in connection with working out offers and executing orders for customers. In connection with making an offer or executing an order, there are a series of specifications which specify the product and how the product is to be produced, assembled, transported, used, serviced and recycled/scrapped. Figure 2.1 illustrates an example of specifications used during a product's life cycle, from the moment when an identified need arises and continuing through design, production, use, service and destruction/recycling.

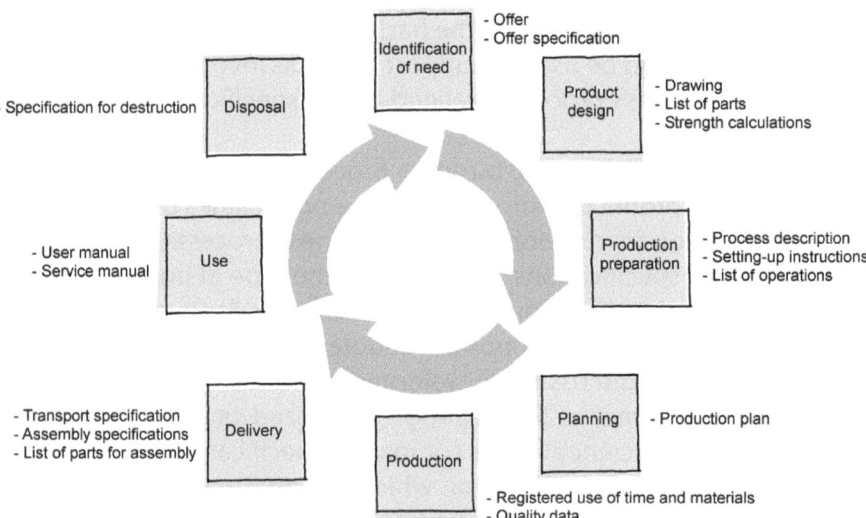

Figure 2.1. *Specifications in the course of a product's life cycle.*

As the figure indicates, many different specifications can describe the product during the course of its life cycle. In the case of a mass-produced product, it is possible to work out all the specifications in connection with development of the product, and subsequently these specifications can be used every time a new product is produced.

If, on the other hand, customer tailored products are manufactured, it will be necessary to work out some of the specifications every time an offer has to be worked out or an order is received for a customized product. We shall from now on use the term specification processes for the activities connected with working out specifications related to a specific customer's needs. Specification processes denote the business processes which analyse the customer's needs, create a product which is adapted to the individual customer, and specify the activities which have to be performed in connection with, for example, purchasing, production, assembly, delivery and servicing of the product concerned (i.e. the product's life cycle properties).

Specification processes

Figure 2.2 shows a number of the activities which are performed as part of a company's specification processes. Those specifications which are worked out form the basis for the subsequent activities: purchasing, planning, production, assembly and delivery.

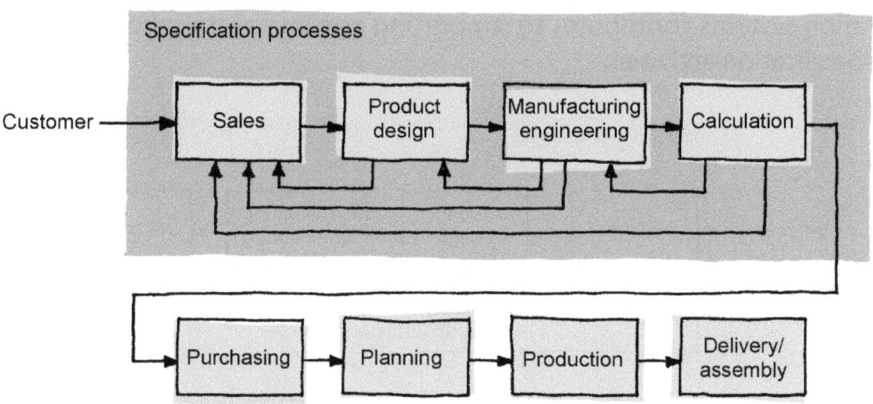

Figure 2.2. *A company's specification processes.*

As can be seen in the figure, the specification activities are typically divided among various departments/functions within the company. Different employees are involved in specifying the product and for example how the product is to be produced, which means that there are many activities which do not create any value, and there are considerable co-ordination problems, both internally in the individual departmemts and between departments.

As described in the example of Doors Inc. in chapter 1, it is now possible to build up configuration systems to support these activities and connect them. An example of this is the use of a product configuration system for sales activities. By building up a product configuration system, engineers engaged in design and manufacturing engineering model the rules for the product's construction, way of working and so on (a so-called product model) so that this knowledge can be expressed explicitly and incorporated into a configuration system, which can subsequently be used by the company's sales staff to configure a product in collaboration with the customer.

This enables the salesman to put together a product which is based on the customer's wishes, and – following up on this – to calculate the price. At the same time, the customer is insured a well-documented description of the product at an early stage in the process.

In principle, the building up of such configuration systems means that knowledge from an organisational unit is modelled and made available for other organisational units. Figure 2.3 shows how building up configuration systems contributes to supporting and integrating the company's specification activities.

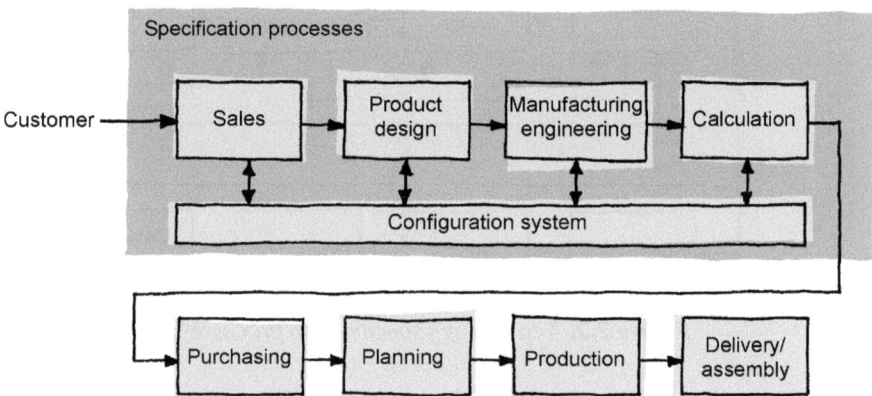

Figure 2.3. *Configuration systems support and integrate the activities in a company's specification processes.*

When knowledge about products and, for example, their production is built into configuration systems, many feedback and coordination problems are avoided. The use of configuration systems leads to radical changes in the specification processes, and thus in the company's internal working procedures and organisation.

The employees' job situation is changed markedly. For the engineers/technicians, the changes mean that whereas they had previously, on the operational level, worked out the specifications for a customized product themselves, they now have to develop and maintain configuration systems, which then support the development of the specifications – or automatically produce the specifications – starting with the customer's wishes. In other words, the engineers and technicians become 'model managers' who not only have to be able, for example, to dimension a concrete product for the individual customer; in order to build up and utilise configuration systems, they also have to be able to express and model their knowledge in a form which can be incorporated into a configuration system.

Specification processes are a part of a company's operational system. In Figure 2.4, the activities in an industrial company are described by means of three activity chains divided up into operations and development, where the operational system is in turn divided into an order chain, covering sales, specification, planning, purchasing and invoicing/post-calculation, and a chain which describes the physical activity of carrying out the production process.

In the upper part of the order chain, a specification of a customized product is made, together with instructions for how it should be produced, assembled, transported, used, serviced and re-cycled (the product's life cycle properties). Production – shown in the production chain – takes place on the basis of the specifications and plans worked out in the order chain.

The specification processes are shown as part of the order chain, which includes the activities performed as part of sales and order execution. Specification of a customized product and its life cycle properties is followed by planning and monitoring production, purchasing of materials, carrying out the production process, invoicing and if necessary post-calculation.

Creating specifications such as bills of materials and operation lists is a pre-requisite for being able to carry out the process of planning production and purchasing. If the company such as Doors Inc. manufactures customized products, then it will normally not be possible to plan the requirements for materials or production capacity before the specifications of bills of materials and operation lists have been worked out, together with the time consumed by each operation.

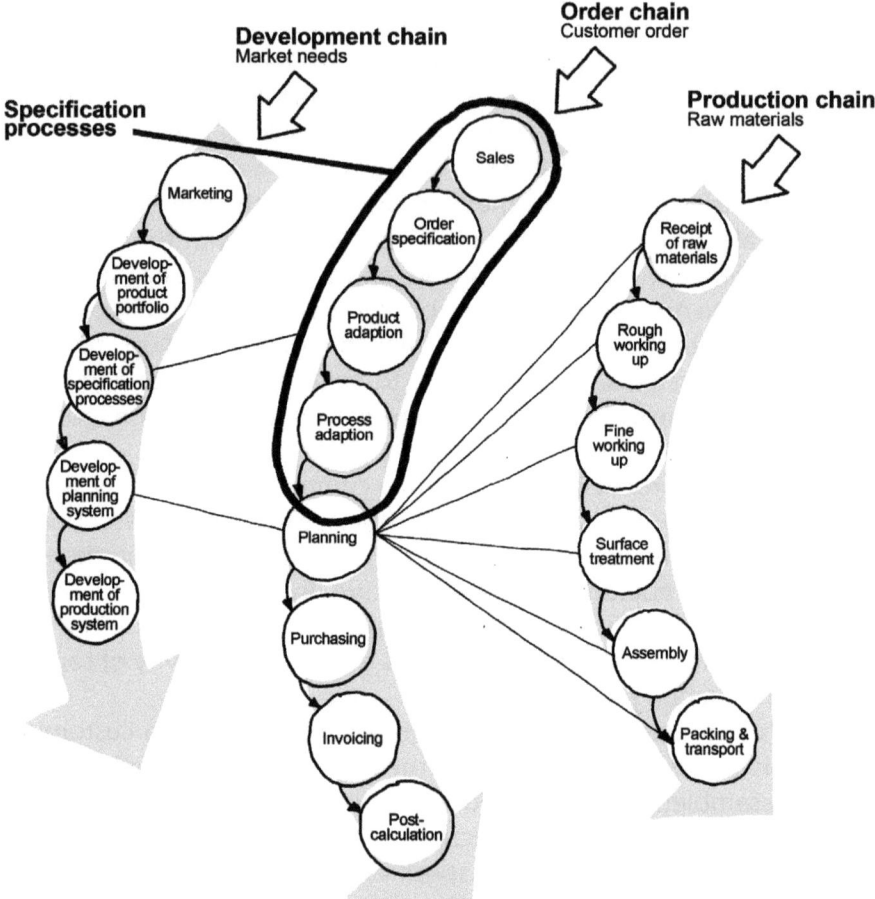

Figure 2.4. *Specification processes and their environment.*

The first chain – the development chain – describes both the development of the range of products and the development of the business processes which lead to a customized product. The figure shows the development of specification processes and configuration systems, together with the development of planning and production systems, as examples of the development of business processes which lead to a customer-specific product. Development of the product range includes the development of modules which can be put together and modified in the order chain – for example by using a configuration system.

In the specification processes in the order chain, the day-to-day specification of customized products and the development of the production specifications etc. are carried out, after which it is possible to perform planning, purchasing, production, assembly, delivery and so on.

In relation to development processes, specification processes can be characterized as having a relatively closed solution space. Thus, the activity of specifying a customized product takes place on the basis of the rules for creating a product variant, which have been described in the development phase. Figure 2.5, taken from [Schwarze, 1996], summarizes some characteristics of development and specification.

Characteristics	Development	Specification
Degree of freedom	High	Low
New modules (components)	Yes	No (pre-defined)
Knowledge	Generated	Utilised/taken into consideration (not generated)
Type of activities	Creative	Routine
Closed-world-assumption	No	Yes

Figure 2.5. Some characteristics of development and specification processes.

Thus, specification of a customized product is a relatively routine activity, which takes place on the basis of the knowledge generated in connection with the development of a product range and, for example, configuration systems. Specification of a customized product ideally takes place via the use of modules developed during the development process. In contrast to this, development processes are creative with an open solution space and a high degree of freedom.

Mass customization and specification processes

In recent years, attention has increasingly been focussed on development of the company's specification processes. This is due to a general tendency towards adapting products so they match the individual customer's current needs. This tendency is expressed in the concept of Mass

Customization [Pine, 1993; Tseng & Piller, 2003; Forza and Salvador, 2007], which states that mass production companies, which have previously produced and sold identical products in large runs, are increasingly dropping this principle and adapting their products to the individual customer. In this way, a new business process arises within the company, with the aim of adapting a product to suit the customer's needs and of creating the specifications which should form the basis for subsequent activities involving production, assembly, delivery and servicing.

A second type of company which to a considerable extent uses configuration systems for supporting customer-oriented specification tasks is companies which manufacture large, complex products. The products can for example be power stations, boiler systems or cement factories, and where the individual customer orders/projects involve a considerable amount of work for designing the individual parts of the plant, so it fulfils the customer's concrete requirements. This type of company, which can be called one-of-a-kind producer, has a considerable challenge in effectivising the task of designing the individual plant/product in detail, so it suits the customer's needs.

To make this task more efficient, it can be an advantage to distinguish between actual development work, in which various parts of the product/ plant are developed from scratch, and the task of designing parts of the product/plant in detail so it matches the individual customer's needs.

The reason for separating the actual task of development from the task of detailed design of a product or plant for the individual customer is that the primary aim of the development task is to develop new creative solutions that increase the general value of the products to the customers and/or reduce costs. On the other hand, the primary aim of detailed design of a product or plant in relation to a concrete customer order is to rapidly and efficiently work out error-free specifications for a product which is adapted to a concrete customer's needs. If the two tasks (development and operations) are mixed together, there will, put colloquially, be too many operational tasks in the development task and too much development work in the customer-oriented projects. An important part of the development task is to develop modules and the associated configuration systems to be used in the operational phase and thus make easier the task of designing a customized product based on a set of modules. This means that the operational phase involves configuring based on a set of modules rather than designing a customized product from scratch.

A third type of enterprise is the large group of companies that manufacture customized products in small series. Doors Inc., which was de-scribed in the introduction, is an example of such a company. For this type of company, the challenge – as for companies engaged in one-off a kind production – is to focus on the customer-oriented business processes concerned with analysing the customer's needs, the creation/adaptation of a product which fulfils the customer's needs, and the creation of the specifications which form the basis for subsequent activities of production, assembly, delivery and servicing.

Figure 2.6 shows the three types of company. For the mass production company moving towards mass customization, adaptation of a product so it fulfils the customer's needs is a new activity that arises. Thus, in this case, it is a question of developing completely new business processes and corresponding configuration systems, which can support the task of specifying customized products. In this type of company, it is of vital importance that the configuration systems which are set up are integrated with the company's ERP-systems, which contain information about products and, for example, their production procedures.

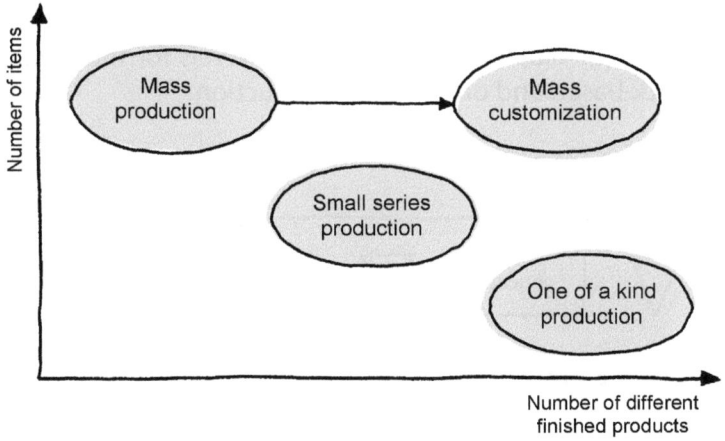

Figure 2.6. *Three main types of industrial companies – and specification processes.*

For companies engaged in one-off a kind or small series production, adaptation/specification of a product to match the customer's needs is not a new process. The challenge in this case is to separate the task of specifying customized products from the actual task of development, and then to formalize and introduce IT-support for the task of customer-oriented specification, so it can be performed in a more rational manner.

For all three types of company, the specification processes comprise the chain of activities which lead to those specifications which are the basis for being able to sell, manufacture, use and dispose of a product. These could for example be specifications describing the product's function, design, manufacture and servicing. A common feature of all three types of companies is that the principles of mass customization mean that a module-based product range is to be built up, and that configuration systems are to be used to support the task of working out specifications in the customer-oriented business processes. An important pre-condition for the company's being able to use modules and configuration systems is that it is possible to develop a product range and a set of business processes that are stable over time. This will normally pre-suppose a focussed market strategy, in which the company chooses which customers it wants to service, and which customers it does not want to sell products to.

Having now described various types of companies with different types of customer-initiated specification processes, we shall now characterize the different types of specification processes, following Hansen [2003].

One basic feature which characterizes production in industrial companies is the dividing line between production for stock and production to order (the Customer Order Decoupling Point). On one side of this line, items are produced for storing, while on the other side they are produced to order. Figure 2.7 illustrates three different levels for the dividing line between stock-based and order-based production.

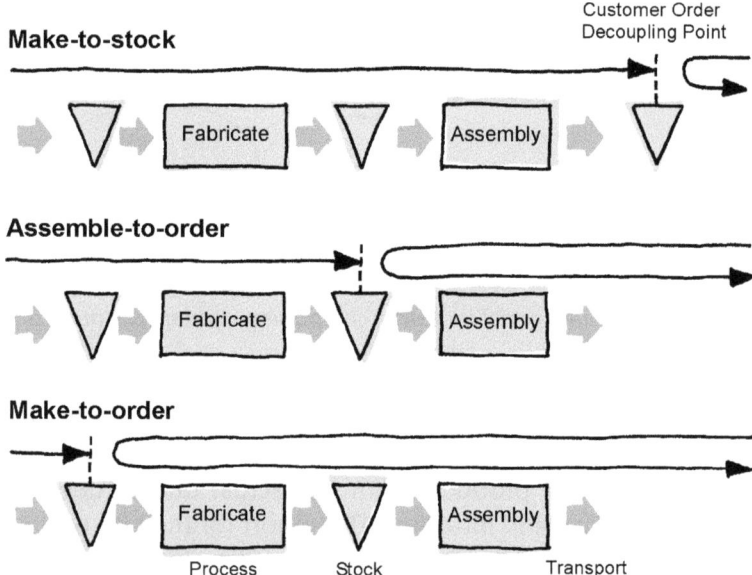

Figure 2.7. The dividing line between stock- and order-based production.

The uppermost level corresponds to a form of production where items are produced to be stored as finished products. The middle level describes a production where items are produced for a component store from which they can be assembled to order. The bottom level corresponds to a form of production where the entire production takes place to order.

In a corresponding way, it is possible to talk about a dividing line for specification processes between order-initiated specifications and specifications which are worked out independently of the individual customer orders (the Customer Order Specification Decoupling Line).

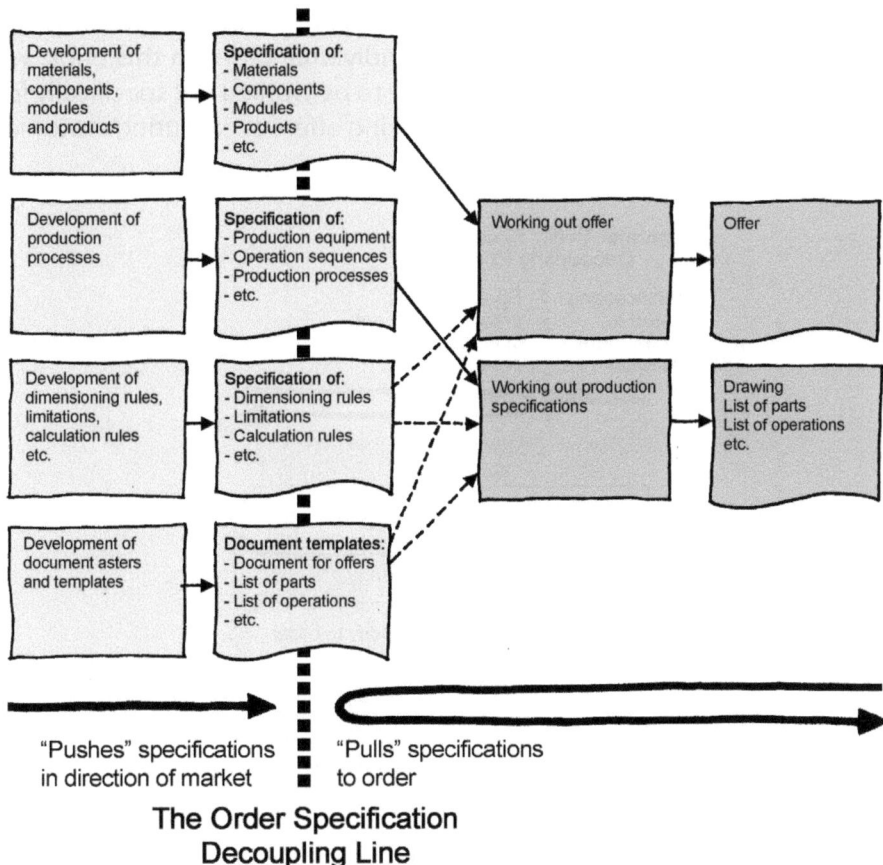

The Order Specification Decoupling Line

Figure 2.8. *The dividing line between specifications worked out on an order-initiated basis and specifications worked out independently of the individual orders.*

Figure 2.8 shows a number of examples of working out specifications. On the left hand side of the dividing line, specifications are worked out independently of the individual customer orders. These are typically specifications, which are the result of developing products, modules or for example production processes. These specifications can be in the form of module descriptions, dimensioning rules, rules for selecting production methods, a set of setting-up instructions which can be used for all products, or a list of parts for a standard component which is being manufactured for stock.

On the right hand side of the dividing line, specifications are worked out for individual orders. These can for example be offers or can take the form of a drawing, a list of parts, a list of operations, a set of assembly instructions, service manuals etc. In other words, specifications can be made independently of the inidvidual customer's order, or specifications can be worked out specifically for the individual order. In this book, we focus on those specifications which have to be worked out specifically for individual orders in connection with making offers or executing customer orders for customized products.

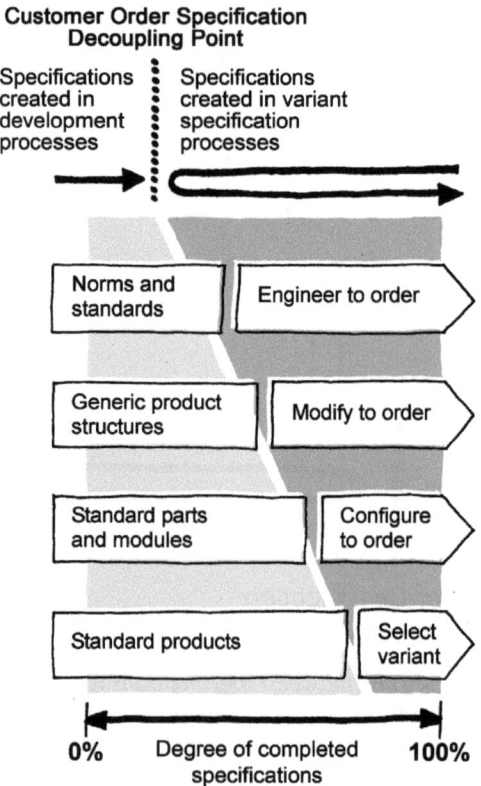

Figure 2.9. *Different types of specification processes [Hansen, 2003].*

The dividing line can help to illustrate different types of specification processes. Figure 2.9 shows four types of specification processes at different positions on the dividing line between specifications worked out on an order-initiated basis and specifications worked out independently of the individual orders (the Customer Order Specification Decoupling Point).

The "Engineer to order" process (the creative specification process) is typical for companies supplying complex products/plants, such as cement factories, spray drying plants or enzyme factories. In this type of company, a considerable amount of work goes into the design and specification of each individual plant. This task is performed using previously designed plants, modules, design manuals and so on as the starting point.

The "Modify to order" process (the flexible specification process) is very similar to the creative specification process and is seen in companies manufacturing customized products. The flexible variant specification process differs from the creative one in that the products are less complex and the creation of a product takes place based on pre-defined modules and using clear sets of rules for how to create a customized product. This means that the specification task is based more on routine and fixed rules.

The "Configure to order" process (the dedicated specification process) describes a specification process where the specifications are worked out automatically by using a configuration system, and where the task of working out the specifications takes place within a finite solution space. This is amongst other things made possible by the use of standard parts and modules, which can be put together in accordance with a set of pre-defined rules.

The process "choice of product variant" denotes a type of specification process in which one chooses a standard product, which to the greatest possible extent fulfils the customer's needs. That is to say, a process in which the seller analyses the customer's needs and, for example by using product catalogues or product databases, selects the product which best matches the customer. The task in this specification process consists in identifying the customer's needs, and then finding the product which best suits the customer in question.

The various types of specification process are described in detail in Hansen [2003]. Companies often have combinations of the types of specification process presented above. The four types have been described to make it possible to analyse and discuss the specification processes which companies work with, and to focus on the positioning of the dividing line between the specifications which are worked out for each customer order and those which are worked out independently of the individual customer order.

Development and use of modules

The use of product configuration is closely connected with the development and use of modules. The aim of developing and using modules is partly to make it possible to create customized products for the market and partly to reduce the number of variants which have to be dealt with internally in the company, and thus to reduce complexity and costs.

Modules are combined by means of a configuration system

Examples of modules could be the engine, the clutch and the gears in a car. The engine module is available in a number of variants, which, for example, can be described in terms of the type of fuel (petrol/diesel), the engine capacity, the control system and the engine suspension. The individual variants of the engine module can be combined with a corresponding number of variants of, for example, the gear and clutch modules, according to a set of rules for which variants of modules it is permissible to use together.

As the example indicates, the number of variants of the individual modules and their possible combinations to form a particular product can rapidly become too large to deal with. In order to handle this complexity, people use a so-called configuration system. This is an expert system that is able to combine modules which are individually described by a number of characteristics, by using rules (constraints) which describe which modules are legal to use in combination. The pre-requisite for being able to develop and use a configuration system is thus that a set of well-defined modules exists with associated rules for how these modules can be combined.

Basically, one can describe a module by saying that it is a limited part of a product with a well-defined function and a well-defined interface to the remaining parts (modules) of the product. For the individual module, the interface to the remaining modules and the rules for how the various modules can be combined are defined. One of the challenges in developing modules is just exactly to develop interfaces to the remaining modules, as these interfaces should preferably remain fixed over a long period of time. At the same time, it should be possible as time goes by to continue to develop the individual modules [Erixon, 1998; Ericsson and Erixon, 2000]. Figure 2.10, from [Pine, 1993], shows some of the main types of modularity.

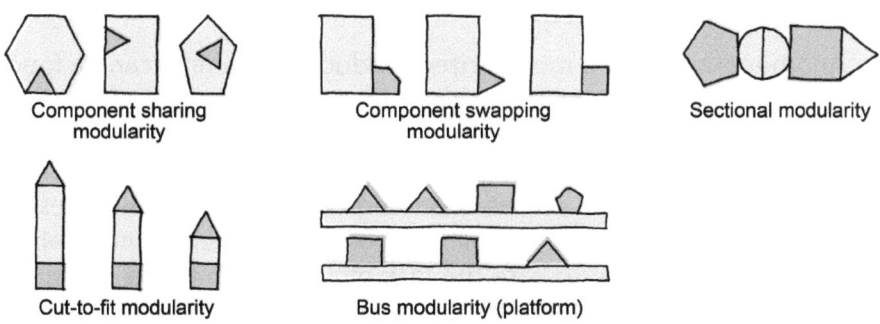

Figure 2.10. *Types of modularity [Pine, 1993, Ulrich & Tung, 1991].*

"Component sharing modularity" and "component swapping modularity" use the same components to span both product variants and product families. With "sharing", the same components are used across product families, as for example in the case of Black & Decker, which use the same type of electric motor in a long series of different tools. With "swapping", variants are introduced into a product family by adding small components; an example of this is "Swatch" watches, where the products are varied by mounting various dials, hands or watchglasses on the same watch mechanism.

"Cut-to-fit" modularity has the property of parametrization, where some of the modules can be adapted by changing their dimensions – for example, clothing that can be adapted to suit the size of the customer. "Sectional modularity" means that modules can be combined freely (like LEGO bricks), by exploiting the modules' interfaces.

"Bus modularity" (or platforms) means that a platform is developed on which components can be mounted. Platforms are used in the computer industry, for example, where PCs have a platform in the form of a mother board, on which it is possible to mount a series of different components such as RAM, processors, graphics cards and so on.

As described in the example of Doors Inc. in the Introduction, when building up a configuration system we want to express knowledge that describes a whole product family. By a product family we mean a group of different products which are created from a common set of components (modules) and which have a number of common characteristics.

An example of a product family could be cars of a particular type, such as the Peugot 407 series. You can get a Peugot 407 in a vast number of variants with different engine capacities, bodywork, trim, accessories etc. The individual variants in a product family are created by using modules for the engine, gears, body and so on.

Use of modules by SCANIA

Another example of a modularized product programme can be found within the company SCANIA, which manufactures trucks. SCANIA has over a considerable number of years developed a modular structure for trucks, with the result that today they can configure a truck to suit the individual customer's needs based on selection and combination of modules. An example of such a module is their gearboxes, which are found in about 2000 different variants. To ease the task of adaptation to the remaining modules, there are only two different gearbox housings. This means that the 2000 different gearboxes only need to be adapted in two ways to the remaining modules such as chassis, engine, clutch and driving axle.

By standardizing the gearbox casing, the lot size increases, which means that it is possible to produce the casings more efficiently. On the other hand, the cost of materials and the weight of the casing may also increase, since in some cases the gearbox casing is somewhat larger than necessary. Thus, the individual gearbox casing may have a number of features which are not quite optimal in relation to the concrete requirements. However, SCANIA uses a standardized gearbox casing anyway, in order to achieve larger production runs and thus more efficient production of gearbox casings, and – perhaps most importantly – standardizing the gearbox casings makes it much easier and thus cheaper to adapt the gearbox casing to the remaining parts of the truck.

The idea can be formulated in the following way: When modules are used, the costs of the individual module rise and, in relation to a number of other factors such as weight or the materials used, a module can appear over-dimensioned or not quite optimal. On the other hand, total costs fall, as the task of combining and adapting the individual modules into a complete product becomes much easier and production becomes simpler and cheaper. In addition, a number of administrative costs are associated with introduction of a new variant within the company, as each product variant has to be set up within the ERP system with information about lists of parts, lists of operations, prices and so on, and this is associated with considerable costs. Such costs are minimized by using products based on modules.

The considerations presented above mean, amongst other things, that one of the biggest challenges when using modules is that the individual employees choose to sub-optimize and develop an individual variant/solution instead of using a standard module. Using modules requires considerable discipline within the company and a marked awareness of the distribution of costs between item costs and total costs.

A number of methods are available for development of modules, for example as described in [Ericsson & Erixon, 2000].

Product models and configuration systems

In the preceding sections of this book, the term configuration system has been used a number of times in the sense of an IT system which can support the task of working out specifications for customer-specific products. We have also stated in connection with product configuration that it is necessary to build up a product model which contains the knowledge about the products that is to be put into the configuration system. We have further mentioned that models exist which not only contain knowledge and information about the products' properties, function and structure, but also models which describe the products' life cycle properties, such as how the products are to be produced, assembled and delivered. In the following section, we describe these concepts in more detail.

Configuration system

The concept of a configuration system arose during the 1980s in connection with the development of a particular form of IT-based knowledge representation known as constraint-based programming. In constraint-based programming, a solution space is defined, such as a series of modules in a product. Then, a number of constraints are defined for how these modules can be combined. The defined constraints are subsequently used to reduce the possible solution space until there is only one solution left – the final configuration. This procedure is carried out by using a so-called inference machine, which together with a knowledge base (containing rules and constraints) and a database, makes up an expert system.

In this book, a product configuration system denotes an IT system that is mainly based on constraint-based programming. In other words, a system which is ideal for combining pre-defined modules in a product in relation to a series of given limitations. The basic principles of expert systems, knowledge representation and reasoning are described in more detail in chapter 7.

To configure means to put together a product from well-defined building blocks (modules) according to a set of pre-defined rules and constraints. Two different definitions of a product configuration system are then:

"Software systems that create, use and maintain product models that allow complete definition of all possible product options and variations with a minimum of entries" [Bourke, 1998, p.1].

"A configurator is a software that assists the person in charge of the configuration task. It is composed of a knowledge base that stores the generic model of the product and a set of assistance tools that help the user finding the solution or selecting components" [Aldanondo et al., 2000].

As the first definition indicates, a configuration system is based on a model of the company's product range, defined as a set of modules and rules for how these modules can be combined. A product model can be considered as an abstract description of a company's product range. Figure 2.11 illustrates the process of going from the real world, which in this case is a set of products and the knowledge which lies behind them, to an IT system, which contains a subset of the knowledge that exists in the real world.

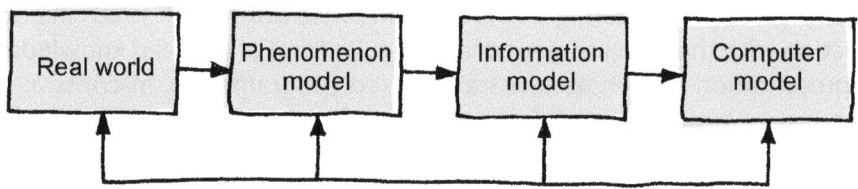

Figure 2.11. *From the real world to an IT system [Duffy et al., 1995].*

The first step is to build up one or more phenomenon models which express relevant aspects of the product range which one wants to incorporate into an IT system. This step corresponds to working out a product variant master, as described in chapters 3 and 5. Next the phenomenon model is formalized, so it can be incorporated into an IT system (corresponding to object oriented modelling), and finally the IT system is programmed.

When the phenomenon to be modelled is a product range, the phenomenon model is known as a product model. The term product model was originally used in the 1970s in connection with the development of CAD systems. In order to operate on a product's geometry, people rapidly discovered that it was necessary to have a basic description of the product. This description was known as a product model. Since then, the concept of a product model has been extended to include all knowledge and information about a product range and the associated life cycle systems.

In the literature, a number of different definitions of the concept of a product model can be found:

"A product model is defined by a total set of characteristics, defining the transformation, function, organ and component structures of a machine system" [Andreasen, 1994].

"A product model is an abstract representation or description, describing (a) the structure of P and (b) facts, objects, concepts and properties that are relevant in any life cycle phase of P. P can be a single product or a family of products. A product is a thing, substance or a service produced by a natural or artificial process" [Schwarze, 1996, p.33].

"A product model is usually intended to define the various data generated through the product life cycle from specification through design to manufacture" [Shaw et al., 1989].

As can be seen from these definitions, a product model is not just a model of the structural form of the products, but also a model of the products' functions and other properties and the way in which the products interact with life cycle systems, for example for production, assembly and so on. Products can be both physical and non-physical (services).

Property models describe the products' properties, for example in relation to the products' strength, weight, centre of gravity and so on. A Finite Element Programme, which performs calculations of products' properties with respect to strength, is an example of a system containing a property model.

In connection with the specification of a product, the product's life cycle – for example design, production, sales, transport, use, service and disposal – is considered to a greater or lesser extent. When the product is specified, the features of its life cycle are to a considerable degree determined. Often it will be necessary to have concrete insight into both the product and the life cycle systems in order to be able to make suitable choices. For example, in order to achieve good transport conditions, it is necessary to know both the transport system and the product.

In the context of building up configuration systems, this means that it often will not be sufficient to have a model of the products, but that larger or smaller parts of the life cycle systems, such as production, transport, use and so on, need to be included in the models. A configuration system that is to support the task of working out production specifications will consist, for example, of a model of the product and models of production and possibly also installation. From now on, these latter models are called "life cycle models".

Delimiting and structuring product knowledge

In connection with building up a configuration system, it is necessary to define some limits for and impose some structure on the knowledge of the product range which is incorporated into the configuration system. As indicated above, it is possible to express many different forms of knowledge about the products' structure, function, properties and life cycle features such as production, assembly, installation, use, service and so on. If you just mix up all these concepts into one huge model, the model will very quickly become quite impossible to comprehend and be impossible to maintain. Thus, it is necessary to impose some kind of structure on the knowledge included in the configuration system.

In order to create a logical structure within the configuration system for the product knowledge incorporated into the system, and in order to systematically define some limits for which knowledge about the product and its life cycle that are to be incorporated into a configuration system, a so-called framework for describing product families is presented.

We can consider a product range from various points of view [Andreasen, 1997]:

- Product structure. We can consider what the products consist of. In other words, how the products are built up and which parts they consist of.
- Product functions and properties. From this point of view, we consider what the products can do (their function) and which properties they have, for example their weight, surface, strength, price and so on.
- Product life cycle properties. A product normally goes through a series of phases from the moment it originates as a need recognised by the customer and further, as it is designed, produced, assembled, transported, installed, used, serviced and disposed of/recycled.
- Variation and family structures. From this point of view, we consider the structures in a product range (modules and platforms) which contribute to creating customized products aimed at the market and at the same time standard solutions/reduced complexity internally within the company.

The framework for modelling product families consists in part of a description of the products in terms of their structure, functions and properties, together with the products' interaction with life cycle systems such as those for production, assembly, installation, use and service. In addition, the framework includes a description of individual products (the instance level) and of a complete product family (generic level).

Figure 2.12 shows the framework used for modelling knowledge about product families. Basically, the framework consists of a model of the structure of the products, models of the products' properties, and models of the products' life cycle, which are called product-related models. Models of the products' properties include for example the products' function or other properties such as weight, surface or strength. Some of the product's properties are formed through the product's interactions with one or more life cycle systems. For example, the cost of manufacturing a product can be calculated by including the product's production and assembly.

The product's function is described under property models. In addition to this, the product's structure and functional units are described under solution principles and part models. The part model describes the components of the product. For example, a bicycle can be described in terms of the parts which together make up a bicycle: frame, wheels, gears etc. In order to produce and assemble a bicycle, it is necessary to have a complete specification of the bicycle's components.

The customer, however, is not particularly interested in the individual components of the bicycle. The customer will typically express his or her needs in terms of demands regarding the bicycle's functions and properties, such as low weight, high stiffness, robustness or high speed. The bicycle dealer then knows that if the customer wants a sports bike with low weight, high stiffness or high speed; then he has to choose a particular combination of frame, wheels and gears to fulfil the customer's requirements.

The relationship between the individual parts of the bicycle and the bicycle's function is expressed via solution principles. A solution principle can be described in terms of functional units. For example, the braking function can be achieved by combining a functional unit which transfers force between the cyclist and the brake pad (the brake cable and handle), together with a functional unit which positions and controls the brake pad in relation to the wheel (the brake pad holder) and a functional unit which transfers a braking force to the wheel (brake pad and wheel). The individual functional units consist of parts. Thus, for example,

	Property models (Derived properties)		Product structre model		Models of the product's meeting with life cycle systems				
	Internal and external properties	Functional properties	Solution principles	Part model	Factory model	Process model	Assembly model	Transport model	Other life cycle models
	Describes consequences of meeting between product and life cycle systems	Describes product's function	Describes product's function-bearing units	Describes product's components	Overall description of production equipment, layout etc.	Detailed description of individual manufacturing processes and production equipment	Describes assembly of product	Describes transport of product	Can for example be service or disposal / recycling
Generic level	Rules for calculating internal properties, e.g. fatigue strength. Rules for calculating external properties, e.g. lifetime.	Rules for describing the product's function and its relation to the function-bearing units (solutions in principle)	Rules for describing solutions in principle and their relation to functions and parts	Rules for describing parts and their relation to solutions in principle	Rules for selection of production equipment and rules for calculating time consumption etc.	Rules for describing the individual production processes	Rules for selecting assembly equipment and calculating assembly time	Rules for selecting form of transport and calculating transport price	
Instance level	e.g. Ultimate strength Product life time Cost price etc.	Functional description	Description / definition of solutions in principle	Drawing, list of parts etc.	List of operations, production layout, description of production equipment etc.	Process description, layout description, description of tools, CNC code etc.	Assembly instructions, list of assembly equipment, assembly time etc.	Transport price, description of packaging, transport documents, etc.	

Figure 2.12. Framework for modelling product families.

the functional unit "transfer of braking force between cyclist and brake pad" consists of a handle, screws for fixing the handle, wire, plastic tubes and so on.

Solutions in principle also express the relationship that a function can be achieved via the use of various different principles. For example, the function of filtering can be achieved by using an electrostatic filter or a mechanical bag filter. The two solution principles for filtering can give the same functionality (filtering), but have radically different part structures.

In the framework, there are also two levels for representation of product knowledge related to single products and product families, respectively. The lower level includes descriptions of individual products (instances), their properties and their life cycle properties. Examples of such instances include drawings, parts lists and instruction manuals for a concrete product. The instances can for example be stored in a database.

The upper level includes generic descriptions of products. In other words, descriptions which describe a whole family of products, and which contain knowledge about how a customized product (instance) can be designed, starting from a customer's needs. A configuration system that can dimension a product, and for example create a letter with an offer, is based on a product model on the generic level containing knowledge about a product family.

In connection with building up a configuration system, it is not normally relevant to include all the sub-models that have been mentioned, but only a subset of them which is relevant in the context of the specification process to be supported by the configuration system.

When developing a configuration system, it is also necessary to consider what degree of detail is to be used. A configuration system used for making offers will normally have a smaller degree of detail than one used to create specifications for carrying out processes such as production or assembly. The configuration system for making offers only contains a partial description of the product, whereas a system used as the basis for working out the specifications for production and assembly normally contains a complete description of all the components in the product.

In chapter 5, "Analysis of product ranges", we give a more detailed description of the various types of models described in the framework. In addition, we describe a number of the analysis tools that can be used to derive the contents of the various sub-models in more detail.

Bibliography

[Aldanondo et al., 2000]: Aldanondo Michel, Guillaume Moynard, Hamou Khaled Hadj; General configurator requirements and modeling elements, Papers from the workshop at ECAI 2000, 14th European conference on artificial intelligence, Humboldt University Berlin, Tyskland, 21-22 august 2000.

[Andreasen, 1994]: Andreasen M. M.; Modelling - The Language of The Designer, Journal of Engineering Design, Vol. 5, No. 2, 1994.

[Andreasen et al., 1997]: Andreasen M. M., C. T. Hansen, N. H. Mortensen; On the identification of product structure laws, 3rd WDK Workshop on Product Structuring, Delft University of Technology, 1997.

[Bourke, 1998]: Bourke R., "Configurators, a status report" article in APICS - The Performance Advantage, May 1998 Issue.

[Duffy & Andreasen, 1995]: Duffy A. H. B., Andreasen M. M.; Enhancing the evolution of design science, Proceedings of ICED 95, Praha, WDK 23, Vol. 1, Heurista, 1995.

[Ericsson & Erixon, 1000]: Ericsson, A. and Erixon, G.: Controlling Design Variants: Modular Product Platforms, American Society of Mechanical Engineers, 2000.

[Erixon, 1998]: Erixon G., "Modular Function Deployment – A Method for Product Modularization", KTH, Stockholm, Doctorial Thesis, 1998.

[Forza & Salvador, 2007]: Forza Cipriano and Salvador Fabrizio; Product Information Management for Mass Customization, Palgrave Macmillan, 2007. ISBN 0-230-00682-5.

[Hansen, 2003]: Hansen Benjamin Loer; Development of Industrial Variant Specification Systems; Ph.D. thesis, Department of Industrial Management and Engineering, Technical University of Denmark, 2003.

[Pine, 1993]: Pine Joseph; "Mass Customization – The new frontier in Business Competition" Harvard Business School Press, Boston 1993.

[Schwarze, 1996]: Schwarze Stephan "Configuration of Multiple-variant Products" BWI, Zürich 1996.

[Shaw & Bloor, 1989]: Shaw N. K., Bloor S.; Product data Models, Research in Engineering Design, 1989.

[Ulrich & Tung, 1991]: Ulrich K., Tung K.; Fundamentals of product modularity; Issues in Design/ Manufacture Integration 1991; American Society of Mechanical Engineers; Design Engineering Division (Publication) DE, 39 p. 73-79, ASME, New York, NY, USA, 1991.

[Tseng et al., 2003]: Tseng Mitchell M. and Piller Frank T. Editors:, "The Customer Centric Enterprise. Advances in Mass Customization and Personalization" Munich 2003, p. 315-328. ISBN 3-540-02492-1.

3

The Procedure

This chapter gives a presentation of the overall procedure for creating, implementing and operating configuration systems. The individual phases in the procedure are described in more detail in the following chapters.

The procedure is intended to help impose some structure on the task of developing, implementing and using configuration systems. By following this procedure, it should be possible to:

- Derive the business requirements for the configuration system to be created, develop the future specification processes, and delimit and define the knowledge which is to be incorporated into a configuration system.

- Analyse and describe a complete product range and the rules for designing a customer-specific product. It should also be possible to evaluate and if necessary adapt the product range so it becomes suitable for configuration.

- Express product knowledge in an appropriate form for incorporation into a standard configuration system.

- Implement the configuration system that has been created, and ensure that the system can be continually maintained and further developed.

In addition, the procedure is intended to help deal with the organisational changes which take place when knowledge about products is formalized and incorporated into a configuration system.

The procedure builds on theories and methods from a number of different technical areas:

- Mass Customization and modularization of products.
- Business Process Reengineering.
- Product development and knowledge about life-cycle systems such as production and assembly.
- Architecture for building product models.
- Modelling techniques such as object-oriented modelling.
- Methods for constructing IT systems, including object-oriented analysis and the use of expert systems.
- Organisational factors related to the construction of configuration systems, including project management and change management.

The procedure presented here builds on larger or smaller parts of the fields mentioned above. The relevant theory is only referred to in this book to the extent that it is a central issue in relation to use of the procedure.

As an introduction to the procedure, we give a short description of a number of so-called project life-cycle models, which describe the progress of a project for developing IT systems. The description includes a short presentation of the historical development of life cycle models from the so-called waterfall model to the cycle and spiral model and on to the object-oriented project life cycle and the Unified Process.

The earliest models for describing the ways in which projects proceed were based on so-called waterfall models, which described a series of development steps to be executed during the development of IT systems. Later, around 1980, people started to describe the progress of an IT project in terms of a cycle, which had to be performed a number of times. A project for development and operation of an IT system normally passes through the following main phases:

- Analysis and initial functional requirement specification.
- Design of solution/high level sketch of the system.
- Programming and test.
- Implementation.
- Maintenance/further development.

The sequence of activities involved in the development and operation of IT systems is shown in figure 3.1 in the form of the so-called project life cycle.

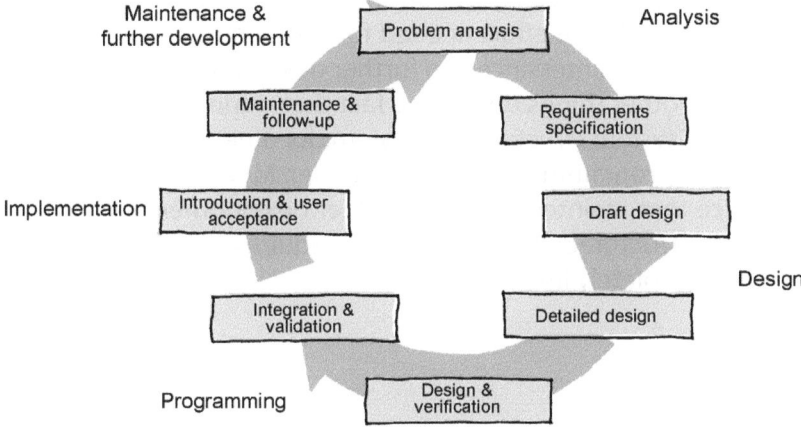

Figure 3.1. *Project life cycle during development of an IT system.*

The project life cycle shown here was described for the first time in the American development project ICAM (Integrated Computer Aided Manufacturing) [ICAM, 1978]. In the ICAM project, in order to support the phases of analysis and design of IT systems, techniques were developed for modelling sequences of activities (IDEF0) and information structure (IDEF1).

At that time, the different phases had different structures and used different notations. In the analysis phase, functional modelling (IDEF0) was used, whereas in the design phase for example a combination of IDEF1 modelling and traditional dataflow diagrams were used, and in the detailed design phase pseudo-programming.

The sequence of activities is shown as a cycle, reflecting the idea that during maintenance and further development of the system a new cycle in the project life cycle is performed, since the same activities are repeated that were performed in developing the first version of the system. The sequence of tasks can thus be performed arbitrarily many times in a given domain, as it is basically the same activities which are to be performed, regardless of whether we are engaged in developing a new system or modifying an existing one.

The reason for introducing the project life cycle was a desire to achieve a more structured procedure for development of IT systems. In the project life cycle, the emphasis is on carrying out analysis and design so that the content and structure of the system is defined and evaluated before the actual task of programming is started.

This reduces the overall costs of development, since the increased effort put into analysis and design, as shown in figure 3.2, usually leads to a marked reduction in the effort needed for programming, implementation, and especially maintenance and further development of the system. This is amongst other things due to the fact that during the analysis and design phases a system structure and system documentation are created, which make the programming task much easier. Moreover, it often turns out in practice that maintenance and further development of an IT system that is not structured and documented requires a lot of resources or may even be completely impossible.

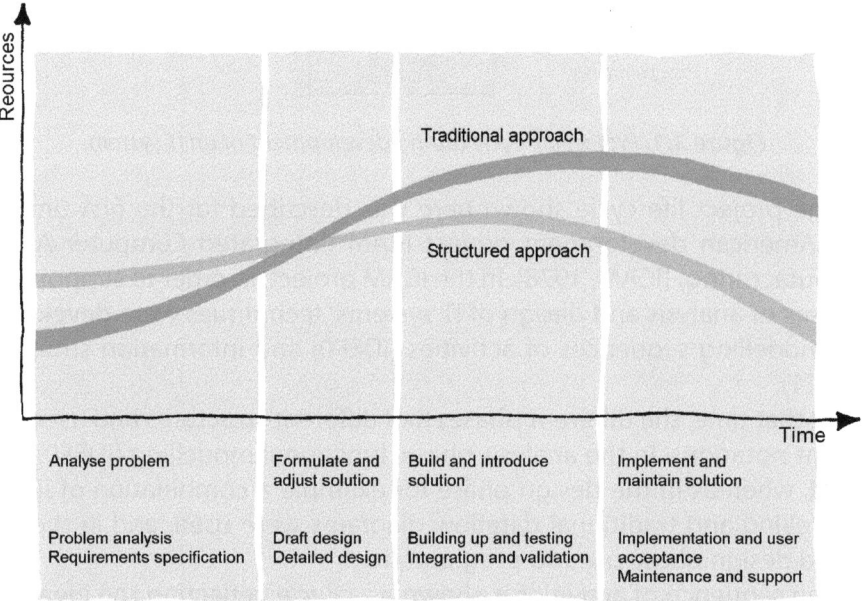

Figure 3.2. *Economies introduced by the use of a structured procedure.*

The project life cycle (from ICAM) described here contributes to structuring and dividing the task of developing IT systems, and covers both the technical and the management aspects of the development task. In other words, the project life cycle both supports the technical activities involved in developing IT systems and works as a tool for managing and organising large development projects, since these can be broken down into series of subactivities with well-defined results, such as the definition of an AS-IS model, definition of a TO-BE model, preliminary design etc.

One of the problems with the early project life-cycle models is that it is difficult to jump between the different phases. At the end of the 1980s, therefore, object-oriented modelling and programming was adopted, as well as the description of project life cycles by means of the so-called spiral model shown in figure 3.3.

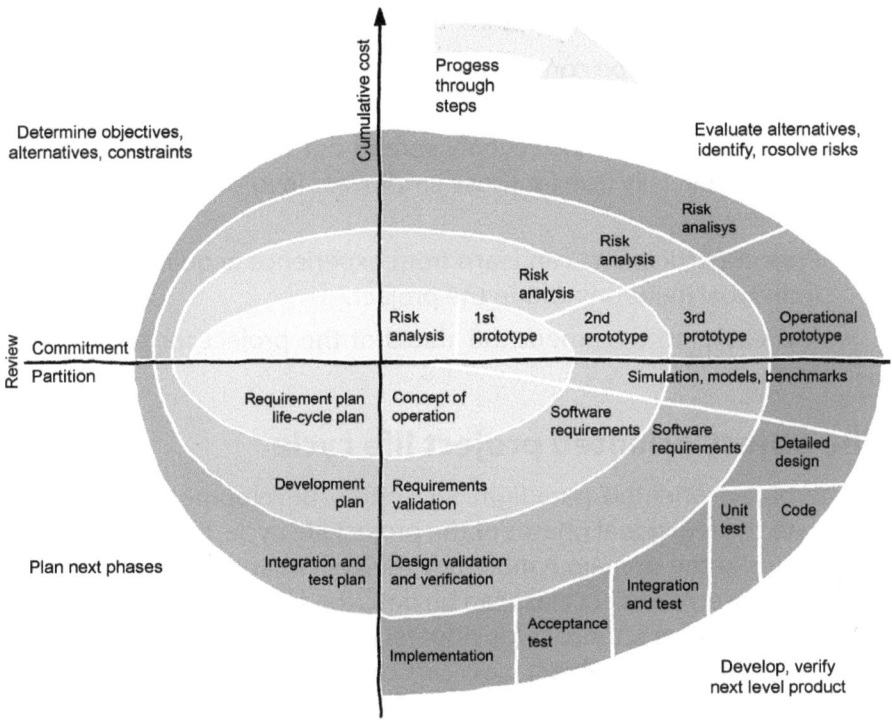

Figure 3.3. *The spiral model [Boehm, 1988], [Bennett et al., 1999].*

In the spiral model, an attempt is made to describe an IT project in terms of a number of executions of the individual project phases, where the IT system is developed gradually via a series of iterations. Use of the spiral model offers a number of advantages in relation to the earlier waterfall model:

- Misunderstandings with respect to the users' needs become apparent early in the process, when it is still possible to change the system.

- The spiral model makes it possible at an early stage to achieve a close dialogue with the coming users of the system, for example by using early prototypes to simulate the users' interaction with the system.

- The spiral model's iterative approach and the closer dialogue with the system's coming users make it easier for the developers to focus on the elements in the system which are most critical for the users, and possibly to leave out less important elements.
- The fact that the system is programmed gradually during the entire course of the project makes it easier to continually evaluate the project's status.
- The system can be continually tested, and errors can be corrected at an early stage in the project.
- The work load is more evenly spread out over the project period. This is especially true for those employees involved in programming and testing.
- Project participants can learn from experience acquired during the individual iterations within the project.
- It is easier to document the status of the project for the project's sponsor.

The object-oriented project life cycle

The object-oriented paradigm for system development attempts to integrate the individual phases of the project life cycle, by identifying object classes in the domain handled by the system at an early stage in the analysis phase. An object class can be described as a collection of objects with common characteristics (attributes) and common behaviour (methods). For example, the object class cars can be considered as a collective term for cars. The class cars has a series of characteristics which identify cars, such as make, motor capacity, weight etc. In addition, the class cars can be associated with a certain behaviour, so that for example the car's annual tax can be calculated as a function of the car's weight and fuel consumption. An object class is represented in IT systems by an independent programme containing attributes (variables) and methods (procedures).

The identified object classes are developed and details are added in all phases of the project life cycle, which actually contains the same phases as ICAM´s project life cycle:

- Analysis.
- Design.
- Development/ implementation.
- Modification/ maintenance.

Figure 3.4 shows the object-oriented project life cycle, which, in comparison with ICAM´s project life cycle, makes it possible to shift more easily between the various phases of system development. This is because (in contrast to previously, where system developers had to change representation between the individual phases) the same structure and basic representation are used in all phases of system development – the same object classes appear in each phase of the project life cycle.

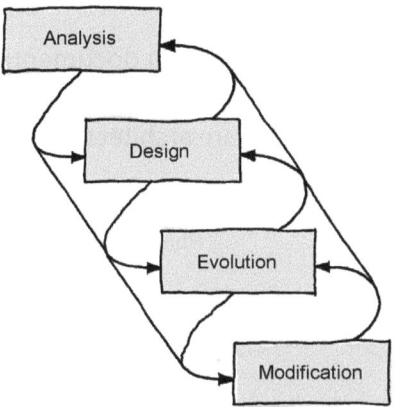

Figure 3.4. The object-oriented project life cycle [Booch, 1991, p. 200].

The overlapping arrows and the arrows leading backward attempt to illustrate how system development based on the object-oriented paradigm provides better possibilities for shifting between the individual phases of the project life cycle. The basic principles of object-oriented modelling are described in more detail later in this chapter and in chapter 6.

Use of the object-oriented project life cycle is also described in "Unified Process", which is the newest project life-cycle model, incorporating most of the experience accumulated in the earlier project life-cycle models. Figure 3.5 shows the content of the individual phases and the weight which the individual activities (workflows) are given in the course of the project's lifetime.

Unified Process only describes the course of actual development. This means that maintenance and support are not included in the model. A number of the most important characteristics of the Unified Process are:

- It is an iterative development process, as in the case of the spiral model.
- The iterative process makes it easier to handle changes in user re-

quirements for the system.

- It relies on the use of object-oriented technologies and the "Unified Modelling Language" (UML) [Booch et al. 1999, www.omg.org] as a notation for use throughout the development process.
- It includes project management as an important focus area.
- It involves continual verification and test of the software.
- Changes in software are managed via systematic use of version management.
- It includes software (CASE-tools) for documenting the system during the development process.
- It includes a choice of software architecture.

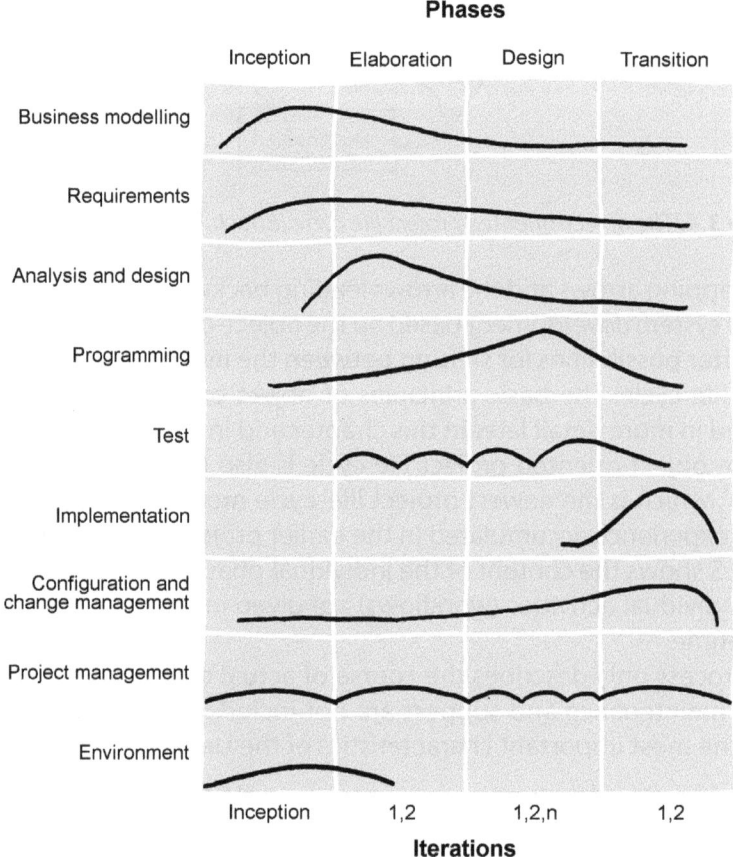

Figure 3.5. *Unified Process [Kruchten, 2000, p. 23].*

In addition, emphasis is placed on working out and continually updating the definition of the system. The system definition (also called the system's scope) includes a concise and precise description of the desired software expressed in natural language.

CASE (Computer Aided Software Engineering) systems are systems which support development and maintenance of IT systems. CASE tools contain, amongst other things, a series of visual tools for showing different parts of the object-oriented model, such as class diagrams, use case diagrams etc. In addition, CASE tools contain facilities for automatic code generation in for example C++ or Java. More modern CASE tools are based on UML and also contain facilities for version management.

The most important requirements for a project life-cycle model can be summarized as:

- It must be possible to carry out iterative development of the system.
- It must be possible to handle changes in the requirements for the system (the system definition).
- Object-oriented technologies must be used for development and documentation of the system.
- Software (CASE tools) must be used for documentation of the system.
- It must be possible to select a software architecture, since this has decisive influence on the functionalities which can be developed.

The procedure for developing configuration systems

Taking the object-oriented project life cycle (Unified Process) as our starting point, we now describe a procedure for developing configuration systems. The procedure contains amongst other things methods for analysing the business processes to be supported by configuration systems, and methods for analysing and modelling a product range.

Figure 3.6 illustrates the procedure for developing and implementing configuration systems.

The first phase in the procedure involves identification and characterization of the most important specification processes, and an analysis of the aims and other requirements for the specification processes to be supported by configuration systems. The specification processes are described and the commercial requirements for the processes are analysed in relation to the company's business strategy.

Phase	Activity	Tools	Results
1 Development of specification processes (Chapter 4)	Step 1. Identification and characterisation of the most important specification processes. Step 2. Formulation of aims and requirements for the individual specification processes. Measurement and gap analysis. Step 3. Design of new specification process. Definition of the configuration system(s) which are to support the specification process. Step 4. Evaluation and selection of scenario. Step 5. Plan of action and organisation of further work.	Flow charts, activity chains or IDEF0 Targeting and gap analysis Framework for structuring product knowledge Other tools: SWOT analysis Scenario techniques Cost-benefit analyses Benchmarking Use case diagrams Project management Change management	Characteristics of the most important specification processes. Aims and requirements for the individual specification processes. Scenarios in the form of descriptions of future specification processes, and definitions of the configuration systems which must support the individual specification processes. Evaluation/ measurement of the individual scenarios' effect. Choice of scenario. Plan of action and plan for organisation of further work.
2 Analysis of product range (Chapter 5)	Analysis of product range. Definition of configuration system's overall content and structure. Design of product variant master.	Product variant master possibly with associated CRC cards Framework for structuring product knowledge Other tools: Modularisation Scenario techniques	Definition of the configuration system's overall content and structure. Product variant master
3 Object-oriented modelling (Chapter 6)	Construction of object-oriented analysis (OOA) model	Class diagram with associated CRC cards and other UML diagrams	OOA model (possibly dynamic). User interface. Requirements specification.
4 Object-oriented design (Chapter 8)	Choice of configuration software. Adaptation of OOA model to the chosen configuration software. Elaboration of requirements specification for programming, including user interface, integration with other systems and programme dynamics	Forms of knowledge representation Criteria for choice of software Class diagram with associated CRC cards Other tools: Other UML diagrams	Choice of configuration software. An adapted OOA model (OOD model). Requirements specification for programming.
5 Programming (Chapter 7)	Programming and test.	Configuration software	A finally programmed configuration system.
6 Implementation	Implementation of configuration system and the future specification process	Plan for implementation Training of users of the system Other tools: Change management	User guide and training plan.
7 Maintenance and further development	Measuring and following up on the new specification process. Maintenance and continual further development of configuration system. Appointment of persons responsible for maintenance and further development.	Measurement methods Plan for organisation of system maintenance	Measurements of the new specification process' performance. Updated OOA model and programme code. People responsible for maintenance and further development.

Figure 3.6. The procedure.

Continuing from this point, scenarios are worked out for the future specification processes, and by using the framework for modelling product families described in chapter 2, the general content of the configuration systems to be developed to support future specification processes is worked out. This forms the basis for the product analysis of phase 2, in which the configuration system's content and structure are described in more detail on the basis of the company's product range, together with the associated life- cycle systems (e.g. production, assembly, delivery, use and after-sales service), if any.

At the end of phase 1, the individual scenarios are evaluated and a scenario is selected for continued development. As a foundation for further work, a budget and an action plan for further work are formulated.

In phase 2, the products are analysed by drawing the product range up in a so-called product variant master. That is to say a description of the products in terms of a specification of the parts/modules included in them and a specification of their functions and properties, together with possible life-cycle properties such as production, assembly and installation. In addition to this, the relations between the individual parts/modules are described.

The aim of this product analysis is to achieve an overall view of the individual product families and their possible variations. In addition, the product variant master helps to ensure that different people in the company have a common view of the product range's structure and possibilities for variation.

For detailed modelling of the IT-systemet itself, object-oriented modelling is used. The product variant master that has been developed serves as the foundation for identification of object classes and class hierarchies. Details of the individual object classes are described by so-called CRC-cards.

The object-oriented model forms a basis for programming the models. In addition, other requirements related to programming of the configuration system, such as its user interface and its integration with other IT systems, such as the company's ERP system, are defined. In phase 4, object-oriented design, configuration software is chosen, and the model that has been developed is adapted to the chosen software.

When the configuration system has been programmed and implemented within the organisation, the system enters an operational phase where it has to be maintained and continually developed further.

Figure 3.7 shows the sequence of activities involved in developing configuration systems. As shown in the figure, the phases of the procedure are normally executed a number of times during construction, implementation and maintenance of configuration systems. Moreover, the process is iterative. There are usually shifts between the individual phases in the process.

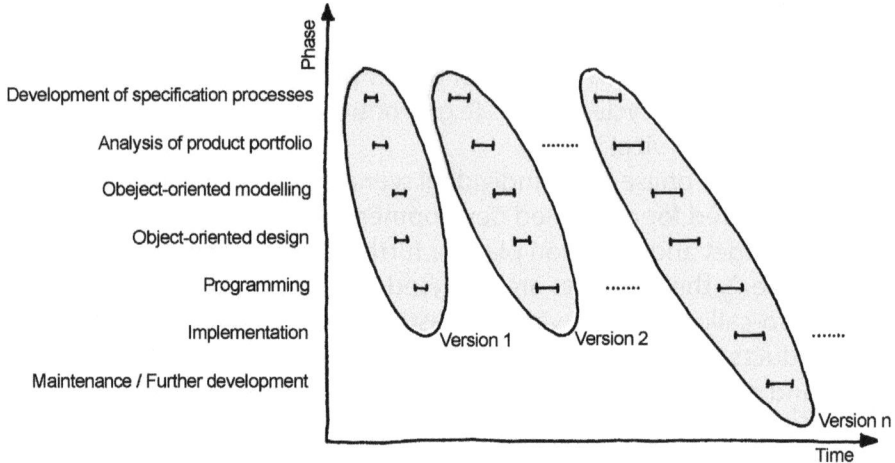

Figure 3.7. *The sequence of activities involved in developing and implementing configuration systems.*

Example of progress of a project

In connection with start-up and execution of the project, it is possible to choose to follow a process in which the phases of the procedure are executed 2 to 3 times. In the first time round – the prototype phase – a rough analysis of the company's specification processes is performed, and preliminary limits are set for the configuration system to be developed. After this, a selected part of the company's product programme is analysed and modelled.

In connection with the prototype phase, the configuration software is selected. The model developed during phase 2-3 can form the basis for testing various standard configuration systems in connection with the choice of software. To complete the first cycle, software is selected, and a prototype is made which forms the basis for the next cycle. In this cycle, a more detailed analysis of the company's specification processes is carried out, and the final limits of the configuration system to be developed are defined. The product variant master and the other masters for the products and life-cycle systems to be included in the system are developed, and the models' class structure and further details are described on the associated CRC cards. Programming of version 2 then takes place.

During the third cycle, the chief emphasis is on test and error correction of the system, prior to the implementation of the system (version 3) within the organisation.

The first cycle, involving making a prototype, is performed rapidly, for example in the course of 1 to 3 weeks. The second cycle typically takes 2 to 4 months, while the third cycle is executed over a period of 1 to 3 months – depending on the size of the project and the number of people involved. The process described above is typical for a company that has not previously developed configuration systems by using a stuctured procedure. If the company is familiar with using configuration systems and has previously carried out projects of this type, it may be possible to omit the prototype phase. In each cycle, a number of iterations are performed, in which parts of the configuration system are gradually developed and tested in relation to users' requirements.

Once version 3 of the system has been implemented in the organisation, the system enters an operational phase, in which emphasis must be placed on continual maintenance and further development. In this connection, it continues to be relevant to execute the individual phases in the procedure and continually analyse the current business requirements for the specification processes and the company's product range. It is also critical that the object-oriented model that has been developed be updated at the same time as the programme is being corrected.

The task of developing and using configuration systems can be divided up into a development phase (phases 1-6) and an operational/maintenance phase (phase 7).

In the development phase, there will be the following roles/tasks:

- The project's sponsor. A person from the company management who has the overall responsibility for the project, including the project's limits with respect to time and money. It is important that the project's sponsor has a certain amount of insight into the technical and commercial problems and opportunities associated with developing configuration systems. It is the sponsor's task to design the use of product configuration into the company's business strategy, and on this foundation, to evaluate the project's relevance and significance for the company. Starting from the project's strategic significance, it is the task of the sponsor to argue for making available the necessary resources for development and operation of the configuration system. It is also the sponsor's responsibility to ensure that the project follows the project plan that has been set up, and that the targets set up are attained. The project's sponsor follows all phases in the project.

- Project leader. A person who has the day-to-day responsibility for the execution of the project. The project leader should know about the procedure for developing configuration systems and have experience from previously performed configuration projects. The project leader takes part in all phases of the project and is responsible for the day-to-day management of the project, including follow-up and, if necessary, correction of the project's work and time plan.

- Facilitator. A person who possesses the necessary knowledge for performing the individual phases of the procedure. This person will typically be a consultant or an employee who has previously carried out projects related to developing configuration systems. If it is the first time that a company develops a configuration system, the facilitator will normally be the only one who has detailed knowledge of how to develop a configuration system. Amongst other things, the facilitator's job is to train the other employees working on the project to be able to contribute to design and implementation of a configuration system. The facilitator typically contributes to the first phases of the project: process analysis, product analysis, and object-oriented analysis and design.

- Change manager. A person who contributes to taking the steps necessary for ensuring that the organisation is ready to perform the project and implement the resulting system. The change manager should have some competence in innovation management in connection with the execution of large technical projects. The task of the change manager is to ensure that employees who are affected by a project for developing and using a configuration system are continually informed and motivated to contribute to the development and operation of the configuration system. It is also the change manager's job to contribute to the solution of any conflicts which may arise.

- The users of the system. The coming users of the configuration system must be involved in the project from its very early phases. It is critical for the users' acceptance and use of the configuration system that they understand how the system is built up, and that they experience the configuration system as a help in their daily work. Especially regarding the users learning to see the system as a help, it is essential that particularly model managers and process managers gain deep insight into the users' work situation and expectations for the system. Such insight can be achieved, for example, by getting together with the coming users of the system and analysing their work situation, or by simulating a series of configuration processes together with the users at an early stage of the development process.

- Model manager. The model manager has the task of collecting knowledge and creating the product variant master, class model, CRC cards etc., to form the foundation for programming the configuration system. In connection with operation of the configuration system, the model manager is responsible for updating the documentation. A model manager can, for example, be a domain expert, who has been trained by the facilitator. The model manager contributes primarily to product analysis (phase 2), detailed modelling in phases 3 and 4, and maintenance and further development (phase 7).

- Process manager. The process manager's task is to develop the company's specification processes and to ensure that the configuration system which is developed gets incorporated into the coming specification process in an appropriate manner. The process manager should have good insight into the company's specification processes and related areas such as sales, planning and production. The process manager contributes primarily to process analysis, together with implementation, maintenance and further development.

- Domain expert. An employee who possesses knowledge of the products and relevant life-cycle systems. It could for example be someone from product development or production. Domain experts contribute with knowledge about parts of the product or the product's life phases, such as production or assembly. Domain experts contribute to process analysis, product analysis and maintenance/further development.

- Programmer. A person who is able to programme the product models that have been developed. The programmer should have detailed knowledge of the configuration software being used, together with knowledge of programming of, for example, customer modules (an application to keep track of e.g. offers made by the configuration system) or integration to other IT systems within the company. If a standard system is being used for implementing the configuration system, then the model manager can perform the programming task – or parts of the programming task. The programmer takes part in the design and programming phases, together with maintenance/ further development of the system.

It must be noted that a single individual can be responsible for several roles/tasks at the same time – for example, the facilitator's and change manager's tasks can be performed by the same person.

In connection with maintenance and further development of the models (phase 7), it is essential that a person (the model manager) be selected to be responsible for maintenance and further development of the models. This person may be able to delegate his responsibility for the updating of rules to various domain experts who possess the necessary knowledge. Phase 7 is critical in the sense that work on developing configuration systems is not finished when phases 1 til 6 have been completed. A configuration system that is not continually updated in order to keep up with development of the products, will rapidly become worthless. The roles in the operation/maintenance phase are described in more detail in the section entitled "Maintenance and further development".

The individual phases in the procedure are now presented in more detail.

Development of specification processes (phase 1)

The first phase of the procedure involves an analysis of the company's specification processes. This analysis is intended to clarify the commercial aims of developing and implementing a configuration system. The analysis must also define the future specification processes and from there go on to define the configuration systems which are to support these specification processes.

In order to clarify the aims of developing and implementing configuration systems to support the company's specification processes, it is necessary to consider the opportunities that can be achieved in relation to the company's overall business strategy. What does it mean for sales, for example, if the time needed for working out an offer and filling an order is reduced from 3 weeks to 30 minutes? Or which strategic significance does it have if production costs are reduced by 10% as a result of better and more error-free specifications in production? Answering these questions requires insight into the company's market position and the customers' requirements.

This book does not go into depth regarding methods for analysis of external conditions and formulation of business strategy. Instead, we describe a number of critical factors in relation to the company's specification processes, such as for example the time required for working out specifications, the possibility of rapidly being able to produce customer-specific products, and the quality of the specifications.

These factors must then be incorporated into the company's strategic planning, since the use of product configuration carries with it a number of opportunities and risks. These have to be clarified in relation to the company's other strategic plans. We refer the reader to the standard literature in this area.

The starting point for the work of phase 1 is identification and characterization of the most important specification processes within the company. This can for example be achieved by identifying the most important specifications, such as offers, lists of parts, lists of operations, or assembly instructions. The specification processes used to produce these specifications can then be described and characterized.

The next step is to formulate the targets and performance requirements for the most important specification processes. In this connection, a series of measurements on the performance of selected specification processes as lead times, resource consumption or specification quality, for example, can be carried out if required. By comparing targets and actual performance within the individual specification processes, a first indication can be obtained of where the greatest potential exists for using a product configuration system to work out specifications.

After having analysed the most important specification processes, scenarios are worked out for how the individual specification processes can be developed from now on by using product configuration. In this connection, the framework for modelling product families presented in chapter 2 is used to produce a definition of the configuration systems to be developed.

A cost/benefit analysis of each scenario is carried out, after which a scenario is selected for further development. Finally, a plan and budget are made for the further work.

In connection with the analysis of the specification processes, it is often necessary to analyse the company's product range. If a company has complex and unstructured specification processes, this is often related to the fact that the product range is complicated and unstructured. To create clarity and structure in the company's specification processes and in the configuration systems which are to support these business processes, it is necessary to obtain an overall view of the company's product range.

Analysis of product range (phase 2)

The aim of phase 2 of the procedure is to provide an overall view of the company's product range, and to describe the product knowledge which is to be incorporated into a configuration system.

A company's product range often appears to be large and have a vast number of variants. To obtain an overall view of the products, the product range is drawn up in a so-called product variant master (cf. figure 3.8).

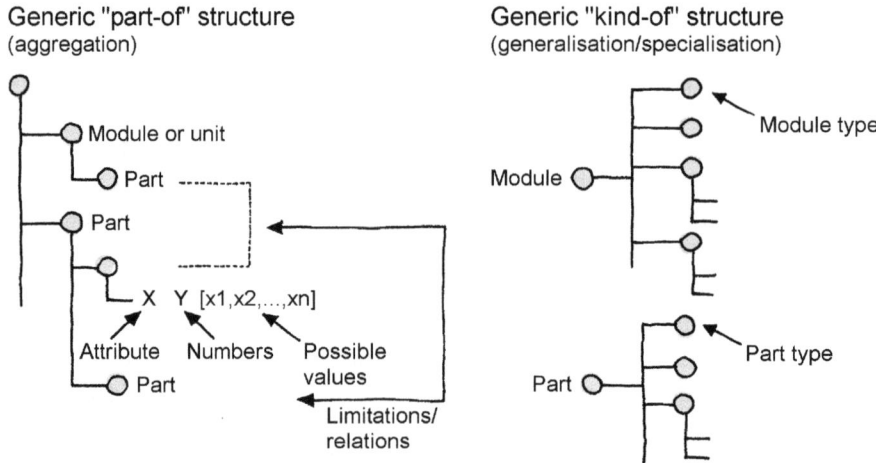

Figure 3.8. *Product variant master.*

A product variant master consists of two parts. The first of these, the "part-of" model (the left-hand side of the product variant master), contains those modules or parts which appear in the entire product family. For example, a car consists of a chassis, motor, brake system etc. Each module/part of the product range is marked with a circle. It is also possible to specify the number of such units used in the product - for example, 1 motor and 4 wheels in each car. The individual modules/parts are also modelled with a series of attributes which describe their properties and characteristics.

The other part of the product variant master (the right-hand side) describes how a product part can appear in several variants. A motor, for example, can be a petrol or diesel motor. This is shown on the product variant master as a generic structure, where the generic part is called the motor, and the specific parts are called petrol motor and diesel motor, respectively.

The two types of structure, "part of" and "kind of", are analogous to the structures of aggregation and specialization within object-oriented modelling.

The individual parts are also described with attributes, as in the part-of model. In the product variant master, a description is also given of the most important connections between modules/parts, i.e. rules for which modules/parts are permitted to be combined. This is done by drawing a line between the two modules/parts and writing the rules which apply for combining the modules/parts concerned. In a similar manner, the life-cycle systems to be modelled are described in terms of masters that for example describe the production system or the assembly system.

The individual modules/parts in the product variant master are further described on so-called CRC cards, which are discussed in more detail in the next section.

It is normally possible to create the product variant master in such a way that the products are described from the customer's point of view at the top, followed by a description of the product range seen from an engineering point of view, while the products are described from a production point of view at the bottom as shown in figure 3.9. This is described in more detail in chapter 5.

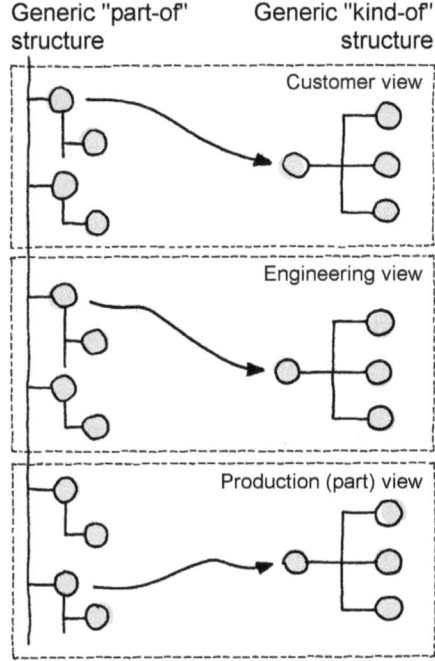

Figure 3.9. *Structure of the product variant master.*

Often, the choice is made to describe life-cycle properties (such as production, assembly and service) on separate masters. In the context of characterizing the individual parts of the product variant master, it is also often advantageous to use tables. This is also the case when rules are given - for example, for how the individual parts of the product can be combined.

The product variant master is drawn up on a large sheet of paper by using for example Visio, Excel or a CAD programme. In connection with the evaluation of the product variant master, there is an opportunity to discuss a number of factors concerning the company's product range:

- Which customer groups do we focus on?
- How well do the various product variants match the relevant customer segments?
- Which properties of the product are "selling points" in relation to the individual market segments?
- Are the right product variants being offered to the market?
- Are all product variants relevant in relation to the market segments being focussed on?
- Are there legal requirements for the product range?
- Which services will we offer the customers, such as manuals, packaging, transport, installation or after-sales service?

The discussion of the company's product range can lead to very important knowledge with respect to how well the product range reaches its target, and where improvements can be introduced in the product range. For example, we might learn that a large proportion of the variants found in the product range have been introduced for internal reasons and can therefore be removed without customers taking any notice. Or maybe in some respects the products fulfil other requirements than those which correspond to the customers' most important needs.

In developing a configuration system, some more technical factors related to the company's product range also need to be explored:

- How can the limits for the individual product families be found? Which market segments and product variants do we want to include in the configuration system?
- How stable is the product range? If changes are regularly being made to the product range's basic structure, it can be difficult to build up a configuration system with a particularly long lifetime. Experience shows that the product range in a great majority of companies has a basic structure that is stable over long periods of time (e.g. 20-50 years).
- How complex is the product? How complicated is it to understand the product's basic technology? If the products are based on complex and inaccessible technology, it is vital that the right people in

the development department, who have an understanding of the products, become part of the configuration team.

- What degree of detail is to be worked with in the configuration system? If the products are large and complex (for example, a cement factory), then it is necessary to base the configuration system on a relatively high abstraction level with few details. A configuration system which is used for making offers often works with a lower degree of detail than a configuration system used, for example, for detail engineering or for working out production specifications.

- How is product knowledge represented? For example, are the products documented in the form of product brochures, drawings, lists of parts, diagrams, and tables? Or is product knowledge mostly to be found solely inside the brains of the engineers?

- Which calculations need to be performed in order to dimension a product variant for a customer? How complex are these calculations? Are the calculations documented? Do the calculations need to be carried out in the configuration system, or are there separate IT tools for performing such calculations and which may have to be integrated with the configuration system?

The product variant master is worked out by domain experts (sales staff, product developers, production staff, purchasers etc.) and model managers, possibly assisted by the facilitator and the project leader through a sequence of for example 6-8 meetings of 1-2 hours each at weekly intervals.

Before the first meeting, the model manager has, perhaps together with 1-2 domain experts, worked out a small part of the product variant master. This work continues at the meeting. During subsequent meetings, work on the product variant master continues so that the individual parts are discussed; when agreement is reached, the result is incorporated into the product variant master or the associated CRC cards.

After the meeting, the electronic version of the product variant master and the CRC cards, if any, are updated. If any information about those parts of the product range which have been discussed at the meeting was not available, it is necessary to assign the individual participants the tasks of collecting supplementary information in order to make final decisions about the product range, The notation and the process of developing a product variant master are discussed further in chapter 5. In the next section, we describe how the information from the product variant master can be further developed into a formal object-oriented class model.

Object-oriented modelling (phase 3)

The aim of phase 3 is to further develop the model from phase 2, which is available in the form of a product variant master and associated CRC cards, into a formal object-oriented model, which contains relevant knowledge about the product range and the configuration system, which can form the basis for the subsequent choice of software, adaptation of the model to match the software, and programming.

In the procedure for developing configuration systems, object-oriented modelling has been chosen as the modelling technique to be used throughout, from constructing the model (product variant master and object-oriented model) to programming and maintenance of the system.

This choice has been made in order to:

- Be able to structure complex knowledge about the company's product range.
- Be able to reuse analysis results throughout the project life cycle from analysis to design, programming, and maintenance/further development of the model.
- Make it possible for domain experts themselves (e.g. product developers and staff in the department for production technique) to model their field of work.
- Facilitate a more appropriate division of work between for example the model manager (domain expert) and system developer (programmer).

An important property of object-oriented analysis is thus the possibility of being able to structure a complex domain by dividing it into topic areas (clusters) and object classes. The chosen structure is used in all phases of the object-oriented project life cycle, which eases the transition between the individual phases, and contributes to a more consistent use of the results produced in each phase.

The fact that the same structural subdivision and notation are used throughout the various phases of the object-oriented life cycle contributes to improved collaboration between model managers, domain experts and programmers, and makes it possible to divide tasks between them. An example is the way that the model manager and domain experts build the object-oriented model, and then the programmer takes over the further task of programming the model.

A further aim of the object-oriented modelling technique is to construct models that are stable with respect to changes, amongst other things by focussing on the most stable elements in the domain and making these the top-level object classes. Finally, an effort is made to exploit the objects' common features through inheritance of the objects' attributes and methods.

The basic principles and notation from Unified Modelling Language (UML) are used. Figure 3.10 illustrates the notation for classes and relationships between classes.

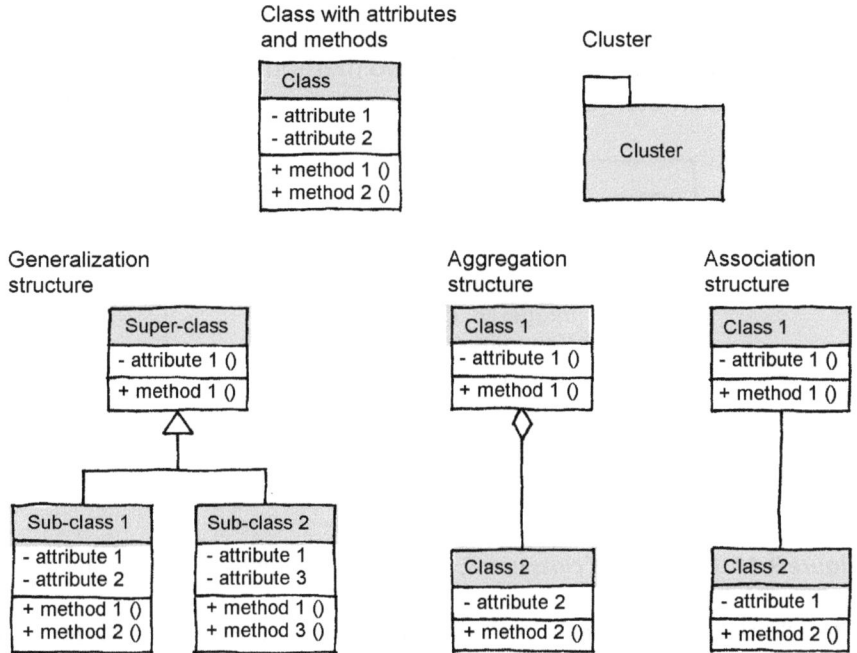

Figure 3.10. *UML notation for defining object classes and their mutual relationships.*

The individual classes are identified by class names, which are representative for the domain being modelled. In relation to a product model, these can for example be names which describe parts of the product, some of the products' functions, or life cycle properties such as production or assembly.

The individual classes contain attributes, which describe what the class "knows", and methods, which describe what the class "does". The attributes can be visible (public) to other classes (marked with +) or protected (private, marked with -), so only the current class' methods can change the value of the attribute. Attributes must as far as possible be kept protected to achieve encapsulation. Methods are normally public, since communication between the classes takes place via methods. Attributes are described by their name, datatype, value interval and units, for example: Length, [Integer], [1..10] mm.

In connection with showing the class diagram, the choice can be made to show only the class name or to include attributes and methods. If methods are included in the class diagram, only the method's name has to be shown. As in the case of attributes, methods can be visible [+] or protected [-]. Figure 3.11 shows an example of a class named "Console" with the attributes Length, Width, Colour, Type and Holediameter_Console. In addition, names are given for two methods: Area_calculation and Check_Holediameter_Console.

Console
- Length [Integer][1..10] mm
- Width [Integer][4..6] mm
- Colour [String][red,green,blue]
- Type [Integer][1,2]
+ Holediameter Console [Integer][2,4] mm
- Area Calculation (): [Integer]
+ Check Holediameter Console (): [Integer]

Figure 3.11. Definition of class name, attributes, and methods in a class diagram.

In addition to this, both attributes and methods are given in detail on the CRC cards, which are described later in the chapter.

There are three different types of relationships between classes: generalization, aggregation and association. In addition, the classes can be grouped in clusters related, for example, to a group of classes describing products or a group describing the production of the products.

Generalization structures describe object classes with common attributes and methods. For example, there could be an object class cars containing general attributes and methods related to cars, such as registration number and motor capacity, while the object class lorries contains attributes and methods specific to lorries, such as maximum load or platform structure. The generalization stucture between the object class cars and the object class lorries signifies that the special object class lorries inherits attributes and methods from the general object class cars. One should as far as possible avoid too large or too deep generalization stuc-tures in the class diagram, as this reduces the transparency of the model. Several smaller hierarchies can be set up instead.

Aggregation structures include object classes, which together make up a whole (for example, the parts of a car, such as body, motor, gear-box, wheels etc). Relationships in an aggregation structure are further described by specifying the cardinality between the two classes - in other words, specification of the number of objects at each end of the relation (for example, one car has four wheels. Object classes in an aggregation structure can communicate with one another, but they do not automati-cally inherit attributes and methods as in the case of a generalization structure.

Association structures describe relationships between object classes that are related to one another without it being a question of object classes which make up a whole. An example could be the object class car and the object class driver. Association structures are described like ag-gregation structures, with a specification of cardinality between the two classes.

In addition to the types of relationships between object classes already mentioned, object classes can be grouped into clusters. A cluster struc-ture indicates a group of object classes, which conceptually form a whole, and which are mutually connected. Thus, a clustered structure normally includes object classes that are mutually connected with generalization, aggregation or association structures. Modelling of object classes and their relationships are described in more detail in chapter 6, Object- Ori-ented Modelling.

In connection with the specification of the product variant master and the final object-oriented class model, CRC cards are used to describe the individual object classes. CRC stands for "Class, Responsibility, and Col-laboration". In other words, this is where a description is made of what defines the class, including the class' name, its possible place in a hierar-chy, together with a date and the name of the person responsible for the class. In addition, the class' task (responsibility), the class' attributes and methods, and which classes it collaborates with (collaboration) are given [Hvam, 1999] [Bellin, 1997] [Hvam et al., 2003].

Figure 3.12 shows an example of a CRC card. In connection with model-ling of products, a sketch of the product part is added, with a specification of the attributes of the product part.

To make sure that we obtain a model that can be understood, it is con-venient to separate knowledge about products and their life-cycle properties (product knowledge) from knowledge about the configuration system's programming (system knowledge), including for example the user interface and integration with other IT systems. In other words, as far as possible, object classes with product knowledge should be grouped in clusters by themselves, and object classes with system knowledge in other clusters.

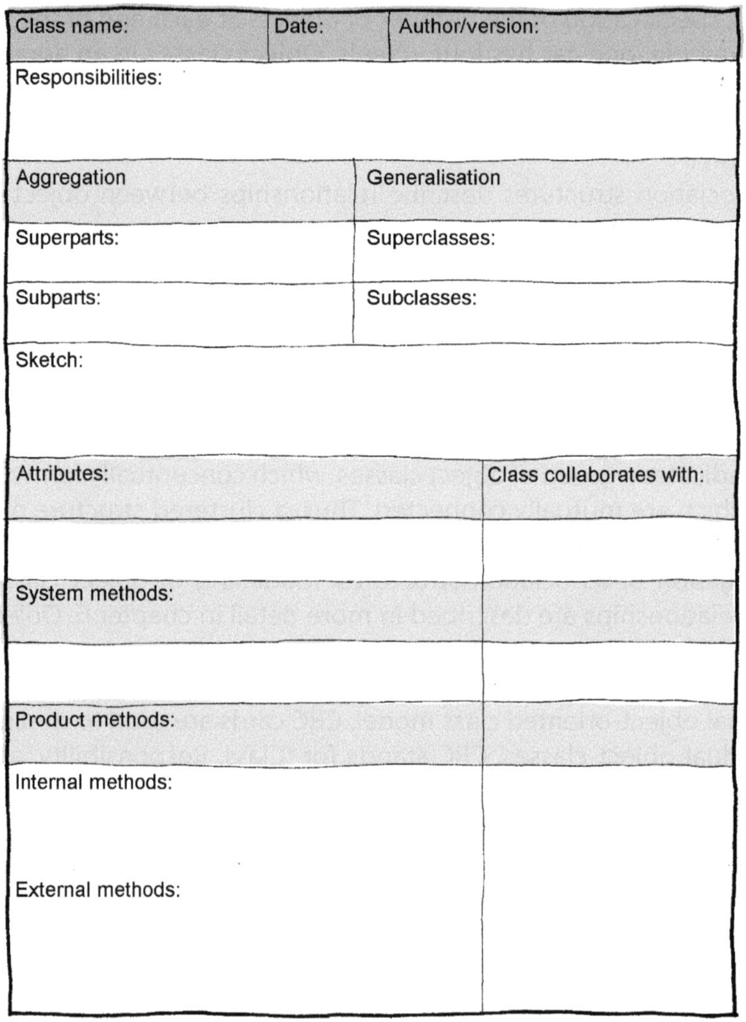

Class name:		Date:	Author/version:
Responsibilities:			

Aggregation	Generalisation
Superparts:	Superclasses:
Subparts:	Subclasses:

Sketch:

Attributes:	Class collaborates with:
System methods:	
Product methods:	
Internal methods:	
External methods:	

Figure 3.12. CRC card.

If an object class should contain both product knowledge and system knowledge, it is possible in connection with the description of methods to divide these up into methods related to product knowledge (product methods) and methods containing system knowledge (system methods). One rationale for this partitioning is amongst other things that system methods do not normally change during the course of the lifetime of the configuration system, whereas product methods need to be updated as products develop and change. Product methods can also be divided into internal methods, which only concern attributes in the class, and external product methods, which require collaboration with other classes.

The CRC cards are filled in gradually during the object-oriented analysis. The CRC cards can be associated with both the product variant master and the OOA model. The purpose of the CRC cards is to document detailed knowledge about attributes and methods for the individual object classes, and to describe the classes' mutual relationships. The CRC cards serve as documentation for both domain experts and system developers, and thus become, together with the product variant master and the class diagram, an important means for communicating and documenting knowledge within the project group.

A more detailed description of the individual fields on the CRC cards is as follows:

Class name

The CRC card is given a name that is unique, so that the class can be identified in the overall structure.

Date

Each card is dated, with the date the card was created, and a date for each time the card is revised.

Author/ version

The person who created the card and/or has the responsibility for revising the card, together with the card's version number.

Responsibilities

This should be a short text describing the mission of the class. This makes it easier to get a rapid general view of what the class does.

Aggregation and generalization

The class' position in relation to other classes is described by specifying the class' place in generalization-specialization structures, respectively, or aggregation structures. This is done by describing which superclasses or subclasses are related to the class within either a generalization-speciali- zation hierarchy or an aggregation hierarchy. Generalization categorises classes with common properties from general to specific classes in the so-called generalization-specialization hierarchies, also known as inherit- ance hierarchies, because the specialized classes further down in the hier- archy "inherit" general properties from the general (higher level) classes.

The other type of aggregation is a structure in which a higher-level class (the whole) consists of a number of subordinate classes (the parts). Using ordinary language, decomposition of an aggregation structure can be expressed by the phrase "has a" and aggregation by "is part of".

Sketch

In working with product descriptions, it is convenient to include a sketch, which in a concise and precise manner describes the attributes in- cluded in the class. Geometric relationsships are usually easier to explain by using a sketch/drawing than by using words.

Class attributes

The various parameters such as height-interval, width-interval etc., which the class knows about itself, are described in the field "Attributes". Attributes are described, as previously mentioned, by their names, da- tatype (if any), range of values and units (for example, Length, [Integer], [1..10] mm). It is often convenient to express attributes in table form.

Class methods

What the class does (for example, calculation of an area) is placed in the field "Methods", which as stated can be dealt up into system methods and product methods.

Methods can be described in natural language with the help of tables, with pseudocode, by using a formal notation such as Object Constraint Language (OCL) [Warmer et al., 1999], which is a standard under UML, or by using the notation from individual items of configuration software. Often, a mixture of the various notations are used. More details of the dif- ferent description methods are presented below.

A method can be described in natural language as in the example be- low:

For signs of type 1 and type 2, a support of type A must be chosen.
For signs of type 3, a support of type B must be chosen.

The same method can also be expressed in table form:

Support type Sign type	Support type A	Support type B
Sign type 1	X	
Sign type 2	X	
Sign type 3		X

Methods can also be expressed in pseudocode, using for example the syntax below:

1. if 'condition' then 'do something'

2. if 'condition' then 'do something' else 'do something else'

3. case 'variable' of
     condition1: do something
     condition2: do something else
     condition3: do a third thing

4. for 'counter variable' = 'start' to 'final value' do
    'do something'

5. while 'counter variable' < 'start' do
    'do something'

The method for selection of support for a sign can then be expressed as:

If Sign type = Sign type 1, Sign type 2

then Support type = Support type A

else Support type = Support type B

A formal notation has also been developed under UML for specifying constraints [Warmer et al., 1999]. The method can be described using OCL as follows:

Context Sign inv:

self.Signtype(1) = self.Support.Supporttype(A)

self.Signtype(2) = self.Support.Supporttype(A)

self.Signtype(3) = self.Support.Supporttype(B)

Finally, it is possible to describe the method by using the notation from a standard item of configuration software:

Sign type = Sign type 1, Sign type 2 → Support type = Support typeA

Sign type = Sign type 3 → Support type = Support type B

Natural language and tables have a low degree of formalization, and can therefore not always describe a method unambiguously. The more formal notations such as pseudocode or OCL are more rigorous and can thus be used to describe a method more unambiguously. In other words, the formal notations have the highest degree of representational strength. On the other hand, they require domain experts to expend the effort of learning the formal notation. Experience has up to now shown that in most projects people choose to describe rules using natural language text, tables and perhaps pseudocode.

Some methods are used in more than one class in the class diagram. In such cases, the method would normally be placed in the class with the greatest affinity to the method. If the method has the same affinity to several classes, then the method should be placed in the class with the highest position in the hierarchy. If the classes are on the same level, then the method can be placed in any one of them. In this last case, it is a good idea to be consistent and, for example, always place the rule in the class to the left. See [Warmer et al., 1999].

Product methods can also, as mentioned, be divided up into internal and external methods. This can be an advantage if module properties are to be described by using CRC cards. In this case, the external methods describe the module's interface to the environment, while the internal methods describe the module's internal structure, functions and properties.

The class collaborates with: (Collaborations)

This field specifies the other classes with which the class must collaborate in order to perform a given action. On the CRC cards, the field is placed in continuation of the class' attributes and methods. This helps explain the relationships between attributes and methods in the different classes.

The structure for the CRC cards shown in figure 3.12 is just one of several possible versions. In some cases, it is relevant to add new fields to the CRC cards, such as references to lists of items in the company's ERP system, references to documents in the company's documentation system, or fields that control who has read or write access.

Use of the CRC cards

Different people take part in the modelling task – for example domain experts, system developers, facilitators and users. It will primarily be the domain experts who have to fill in the CRC cards, since it is they who are in possession of the necessary technical knowledge about the products, processes etc.

It is important not to put too much detail into the CRC cards in the first instance. It is time-consuming and one can easily lose one's overall view of things. Thus, it can be an advantage to start by only filling in the fields "Class name" and "Responsibility". The remaining fields are then filled in gradually, as insight into the structure of the product model increases.

If the CRC cards are to be used for running documentation, it is an advantage if the CRC cards are available in electronic form or are integrated into the actual configuration system's programme code.

Many resources are needed if CRC cards are to be maintained manually as running documentation for the product model. If the software makes it possible to document the programme, for example in the form of notes or text files associated with the individual objects, then this is in some cases sufficient documentation. The problem however is that these text files do not normally describe the programme structure well enough. Important items of information such as superparts/subparts, superclass/subclass, sketch, responsibilities and collaborations can be difficult to incorporate into such notes/text files.

Some companies have elected to place the documentation in their internal file-sharing system, using for example LotusNotes, where the CRC cards are stored in a file structure that corresponds to the structure of the product variant master or the class diagram.

The process of building up the OOA model

The object oriented model is built up through an iterative process in which use is made of use case diagrams (described in chapter 6), class diagrams and CRC cards. Use case diagrams are used initially in order to obtain a better understanding of the users' requirements for the configuration system, and the requirements for integration with other IT systems.

The specified product variant master forms the basis for defining the object classes and class structures in the product model. The individual nodes in the product variant master are candidates for object classes, while part-of and kind-of structures can be recognised as structures between the object classes.

In the product variant master, so-called part-of and kind-of structures are used. These concepts are taken from object-oriented modelling, as described above. Thus, the part-of structure in the product variant master corresponds to an aggregation structure, while the kind-of structure corresponds to a generalization structure.

Figure 3.13 illustrates how a "part-of" structure in the product variant master is transferred to an aggregation structure in the class diagram. Note that the node "Clear plastic film" has not been created as a class. This is because we have chosen to create Clear plastic film as an attribute in the superior class "Console module".

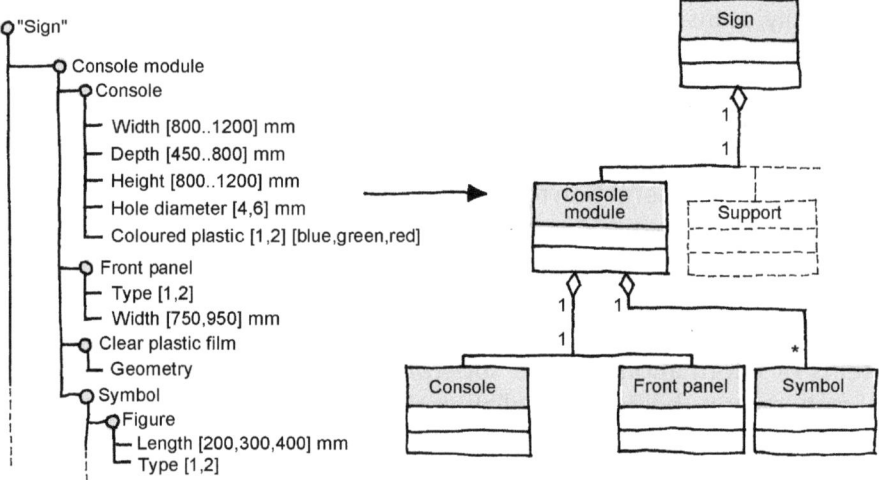

Figure 3.13. *From "part-of" structure to aggregation structure.*

Correspondingly, figure 3.14 shows how a "kind-of" structure in the product variant master can be transferred to a generalization structure in the class diagram.

Not all nodes in the product variant master have to be created automatically as classes. One can for example choose to create some of them as attributes in the superior class. In addition, the choice is often to structure the class diagram in topic-related layers (clusters), in order to obtain a structure that is easier to comprehend. In this connection, one can choose to give the class model a structure derived from the concepts described in connection with the framework for modelling product families presented in chapter 2, for example, in terms of the products' functions/ properties, solution principles/-function-carrying units, components or life cycle properties.

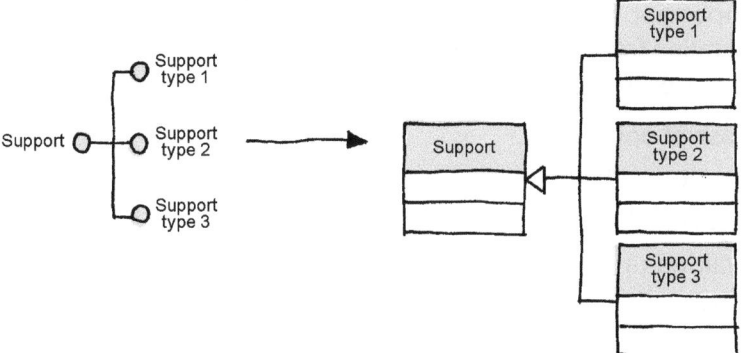

Figure 3.14. From"kind of" structure to generalization structure.

System knowledge, which for example includes a description of how the system should be integrated into other IT systems, will be defined during phases 3 to 5: object-oriented analysis, design and programming.

The modelling task in phase 3 is performed by the model manager, with support from the technical facilitator and in collaboration with the domain experts and the person(s) who subsequently should programme the configuration system.

In the context of developing and filling out details in the object-oriented model, it is important that the model managers and the domain experts are given ownership of the model that is developed. It is vital for the operation and maintenance of the configuration system that the model managers and domain experts are able to update the model and the configuration system, as the products develop further.

It may also be the case that engaging model managers and domain experts in modelling their knowledge about products can lead to many useful ideas and changes in the product range. When domain experts model their knowledge of products, the product structure becomes clarified and re-evaluated. This may also be true in other areas, such as production or after-sales service.

Product models are often developed by a few people from the company collecting information from the domain experts, mostly through meetings or interviews. In the first place, this can lead to errors in the information collected, and secondly there is a risk that the domain experts do not feel any affiliation with the product model. This lack of affiliation can lead to problems when the product model has to be maintained or developed further. Thus, it should primarily be the domain experts who participate in design and maintenance.

In connection with designing the configuration system's user interface, it is important to ensure that the system is user-friendly. The following points, taken from [Rogoll and Piller, 2004], are of vital significance for the system's user-friendliness:

- Ease of operation
- Ease of navigation
- Individual access to information
- Waiting time
- Support

Ease of operation means that it is easy to gain an overall understanding of the system and that it is intuitively easy to approach. One way this can be achieved is by using standard symbols and tools that the user knows in advance. It is also possible to use hints and texts that explain to the user why a configuration is not permitted. The system's configuration steps should be logically structured so that the user can always know where he is in the process and how to continue from there.

Ease of navigation means that it is easy to orientate yourself and "find your way around" as a user of the configuration system. One way this can be achieved is by continually showing the user where he is in the configuration system, possibly in graphic form via the use of images or drawings. In addition, the user should be able to trace his progress through the system and easily go back and alter earlier choices.

Individual access to information. Users of a configuration system are different and have different backgrounds. This can be counteracted by studying the system's future users and training them in the competences necessary for using the configuration system. In addition, information must be made available in a number of different ways, so that different users' needs are taken into consideration.

Waiting time. It is important that the configuration system has short response times, since long response times stress the users of the system and lead to a less coherent configuration process.

Support. It is important for the users' perception of the configuration system that they can always receive support to solve any problems that may arise. This can be done partly by building support and help functions into the configuration system, and partly by having one or more employees with extensive knowledge of the configuration system, who can be contacted and give support when users need it.

In addition to the actual object-oriented class model and user interface, definitions of the system's requirements specification, dynamics, and integration with other systems are produced. These are described in more detail in chapter 6.

Object-oriented design (phase 4)

The phase of object-oriented design involves selection of software, adaptation of the object-oriented model to the chosen software, and definition of the requirements specification for programming.

Choice of configuration software

The starting point for selecting configuration software is the system's requirements specification, which, in addition to the model which has been developed, involves a series of requirements for the system and the supplier, such as the possibility of integrating other systems, response time, operational reliability, and the possibility of support from the supplier. In chapter 8, we present a more detailed description of some of these factors.

When formulating requirements for configuration software, it is important to allocate priorities to the requirements regarding which requirements are essential for the execution of the project ("need to have") and which are of less significance ("nice to have").

The software is selected on the basis of the requirements specification which has been produced. In the first phase, the market is scanned in order to identify maybe 3-5 suppliers which directly appear to be most capable of fulfilling the requirements. For this initial selection, one can for example use the information about configuration systems which can be found on the web page www.productmodels.org/ Configurator-overview.

Then, the final selection of software can be made through a process of collecting more information about the suppliers' software as well as requests to the suppliers to implement a selected part of the product model in their software, so that the system can be tested in practice.

Selection thus takes place on the basis of a list of questions to which the supplier is asked to give a binding response, and a trial exercise, which the supplier is asked to implement in his configuration software. Normally, after agreeing with the suppliers on what is going to be done, they receive a period of 2-4 weeks to answer the questions and to programme the trial exercise.

When evaluating the suppliers' response to the questions concerning the requirements specification and the trial exercise, the most important criteria are normally whether the supplier can guarantee a solution satisfying the most important requirements in the requirements specification; whether the proposed solution is based on standard software ("out of the box technology"); and whether the supplier can give a maximum guarantee that the specified product model can be implemented within the project's planned timespan.

Adaptation of the product model to the chosen software

The object-oriented modelling technique is in principle independent of which software is chosen for implementation of the system, and the model which has been developed can thus be used as documentation, irrespective of the choice of implementation tool. In practice however, it is often necessary to make a number of adjustments to the model before it can be implemented in a concrete item of configuration software.

The standard systems which are currently available for product configuration are not all fully object-oriented. Thus, it is often necessary to adapt the object-oriented model to the possibilities existing for structuring models in the concrete configuration software. In this connection, it must be decided how to document any differences between the structure in the specified product model and the structure in the configuration system.

The adjusted object-oriented model is known as an object-oriented design model (OOD). Apart from this, a detailed specification of the user interface, dynamics and integration with other systems is produced, and set in relation to the chosen configuration software.

Summary of system requirements

Before programming of the configuration system is initiated, the requirements for the system, based on the class diagram, the CRC cards, the definition of the user interface, use case diagrams etc., are summarized and clarified. In addition, requirements for integration with other systems, and for data management and performance (e.g. response times) are specified.

The purpose of the requirements specification is to provide a picture of what is needed in the system which is complete, consistent and usable in the programming phase. The requirements are partly the same as for the choice of software. A number of the factors that can be included in the requirements specification are:

- *User-friendliness*: It must be specificed how user-friendly the system must be. Is anyone supposed to be able to use it without having any previous knowledge of the domain or the system, or is the system to be set up for expert users?
- *Reliability*: How reliable must the system be? What are the consequences if the system fails? Is there any subsequent control of the results arrived at by the system?
- *Accessibility*: How accessible must the system be? Which requirements are there for security? May the system "crash"? How many users will the system have? How often will it be used?
- *Ease of maintenance*: Is it important that the system is easy to maintain? Who is supposed to be able to maintain the system? How often will it be relevant to do so?
- *Performance*: How high a performance must the system have? What are the requirements for response time and number of simultaneous users?
- *Interfaces*: What requirements are there for interfaces to other systems? Must there be on-line access to the external systems, or is file transfer between the systems sufficient?
- *Software environment*: Which operating system is the system to work under? Is the system to be Web-based? What other properties must the system have?
- *Documentation*: What are the requirements for maintenance of system documentation? How often must the documentation be updated? How many different people are involved updating system documentation?

In addition, it is necessary to consider which resources are available for system development:

- How long may the development take, at most?
- What may the system cost?
- How much manpower is available?
- What qualifications must the system developers have?

These are just envisaged as a number of factors that can be considered when working out the requirements specification during the design phase. Chapter 8, deals with the task of working out the requirements specification in more detail.

Programming (phase 5)

The object-oriented model, with the class diagram, CRC card, description of the user interface etc., form the basis for programming the configuration system.

An object-oriented model can in principle be programmed in both an object-oriented and a non-object-oriented programming language. Graham emphasizes the following advantages of using an object-oriented programming language [Graham, 1991]:

- It becomes possible to re-use previously developed codes.
- There is a better possibility of extending existing programmes.
- It supports a well-defined conceptual apparatus, which can be used for both analysis and programming. In this way it becomes possible to model complex relationships. A structure is imposed on the tasks of analysis, design and programming, and finally there will be an easier and more direct transition between the individual phases in the development process. Furthermore, it makes for easier maintenance of the system.

It should be noted in connection with the first two points above that developments in object-oriented programming languages are moving very fast. As a result, the current programming languages can come to be replaced by new languages that are not necessarily compatible with the existing ones, so that existing codes and the associated class libraries cannot directly be re-used.

Another important aspect of the use of object-oriented programming is that if one uses an object-oriented programming language, then the design model is a sufficient foundation for programming and can be used directly as documentation [Booch, 1991].

Often one will choose to implement the models in a standard configuration system. Most of the standard configuration systems available on the market today are knowledge-based systems, i.e. they contain a knowledge base and an inference machine. These concepts are further described in chapter 7.

The programmer should have an in-depth understanding of the standard system to be used. As stated, it is often possible to train model managers or domain experts to carry out parts of the programming task. However, experience shows that it can be difficult for a non-programmer to programme complex rules and associated applications – such as customer modules or integration with other systems – in an efficient manner.

During programming, it is necessary to programme and test the most critical parts of the system at an early stage. It is also necessary to carry out tests of small parts of the system as programming progresses. If too large a part of the system is tested at one time, then it becomes difficult and unpredictable to find errors.

The users (or selected users) of the configuration system should be involved in the test of prototypes and of critical parts of the configuration system. In this way, two things are achieved: First, the users are informed at an early stage about how the system works; and second, input is received from the users at an early stage about desired changes in the system. Thus, the time needed for making changes in the configuration system is reduced and the users are more satisfied.

In the context of programming, it should also be noted that the way in which one chooses to programme classes, attributes and methods can be critical for the system's performance. One should for example be aware that the assignment of attribute values can affect the size of the solution space and thus the system's reponse time. It can therefore on some occasions be necessary to proceed by trial and error, and to programme the same knowledge in several ways until the right solution is discovered.

Programming in a non object-oriented configuration system

If the configuration system is not object-oriented, then the structure in the product model that has been developed cannot directly be implemented in the configuration system. A configuration system that is not object-oriented does not for example support encapsulation of object classes, relations between object classes, or inheritance. Instead, attributes and methods are put into a "flat" structure of folders.

On the basis of the class model, the structure of the folders containing attributes (defined as data types such as "Boolean", "Identifier", "Integer Real", "String" etc.) and their possible variations is defined. In the same manner, a corresponding folder structure is created with methods, which can for example be implemented in the form of resources and calculations.

There are currently a number of standard configuration systems that support the use of tables for representing rules within the system. The use of tables for representing rules makes the implementation task much easier. It is also much easier for model managers, domain experts and programmers to communicate with one another if tables are used, amongst other things because part of the knowledge about the company's product range is in many cases already represented in the form of tables.

Programming in an object-oriented configuration system

If the configuration system is object-oriented, then the class model that has been developed is implemented directly within the system. An object-oriented expert system is based on the principles in object-oriented modelling, and can thus deal with encapsulation of object classes, inheritance and relationships between object classes. During programming of an object- oriented configuration system, the class structure is set up first, after which attributes and methods are programmed together for each object class.

Use of an object-oriented configuration system makes it easier to impose a structure on the knowledge base, which in turn makes it much easier to maintain and further develop the configuration system.

In connection with programming a configuration system, it is important continually to test the individual parts of the configuration system, firstly to find any errors at as early a stage as possible, and secondly to test whether it is at all possible to implement the individual parts of the system. The selected structure of classes, objects, attributes and methods is essential for the configuration system's performance, for example in relation to response times. Thus, it can often be necessary to programme the same part of the system in different ways and then choose the structure which demonstrates the best performance on the basis of tests.

It is also important to carry out user tests of the system and the system's user interface, starting with the very earliest versions of the configuration system. This is partly to ensure input from the users, and partly to ensure that coming users will accept the system.

Implementation (phase 6)

In connection with implementation of the system, it is of decisive importance that the system's users accept the system. Below, we list a number of steps that can contribute to the users' acceptance of the system:

- Ensure that the configuration system is user-friendly and supports relevant tasks for the user.
- Involve the users at an early stage of the project, for example in connection with discussion of the user interface and the requirements

for the system.

- Involve selected users in test of prototypes and early versions of the system.
- Motivate the users, for example by continually giving them information about the new system and the commercial advantages expected from it.
- Give everyone in the organisation information about the project and the organisational changes which use of the product configuration system will bring.
- Give a clear description of the users' future work situation.
- Train the users in the products, in use of the configuration system, and in the future specification processes. It is important that the users of the system (who are often sales staff) understand the choices made by the configuration system, so they feel at ease with the system.
- Introduce monitoring and salary systems or discount systems, which reward users who use the system efficiently.

In connection with implementation of a configuration system, it is important that the system is thoroughly tested and fault-free before it is put into operation. If the users of the configuration system get the idea that the system is unstable or full of errors, they will rapidly lose confidence in the system.

It is also necessary to consider how to implement the system. If using a configuration system is something new for the company, then it could be a good idea to implement the system gradually by first introducing the system to selected customers and sales staff. When these employees and customers have gained experience with the system and are satisfied, then the circle of users can be gradually extended. In this way, a circle of "supporters" can be built up who back the system up.

Maintenance and further development (phase 7)

The configuration system is documented by using the product variant master, class diagrams, and CRC cards. The system can be documented further, for example by using flow charts, use case diagrams, descriptions of user interfaces etc. The core of the documentation is the pieces of information which are to be found in the product variant master, class diagram and CRC cards.

In the context of system maintenance, it is often enough to make use of either *the product variant master with associated CRC cards* or *class diagrams with associated CRC cards*. As the product variant master and the class diagram partly contain the same information, one should choose whether to use the product variant master or the class diagram for the running documentation. The two approaches to documentation of the system are illustrated below.

Figure 3.15 illustrates how the product variant master with associated CRC cards are used for documenting the system.

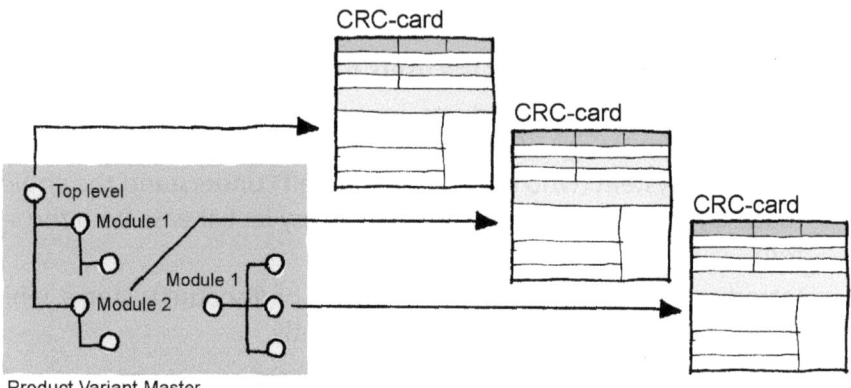

Figure 3.15. *Product variant master and CRC cards as documentation.*

The company GEA-Niro uses the product variant master and CRC cards for documentation of their configuration systems. The company has chosen to incorporate the product variant master and the CRC cards in a Lotus Note application. The product variant master and the CRC cards do not incorporate knowledge of the architecture of the IT system etc., so this knowledge is documented separately.

Figure 3.16 illustrates how the configuration system is documented using a class diagram and CRC cards.

The class diagrams have a more rigorous and formal notation than the product variant master. With respect to the documentation of software, it is therefore a good idea to document the system by using the class diagram and associated CRC cards. On the other hand, domain experts often find it easier to understand the product variant master, as its notation is closer to the concepts and structures the domain experts are familiar with in their daily work.

It is thus a question of weighing whether to use a class diagram or a product variant master in the running documentation. F.L. Smidth uses the product variant master with associated CRC cards for documentation for domain experts, whereas the class diagram and CRC cards are used by programme developers for documenting software. Thus, the same CRC cards are associated with both the class diagram and the product variant master. This solution was chosen because the selected configuration software is fully object-oriented, and because the domain experts do not wish to communicate using a formal IT notation. F.L. Smidth's configuration project is described in more detail in chapter 9.

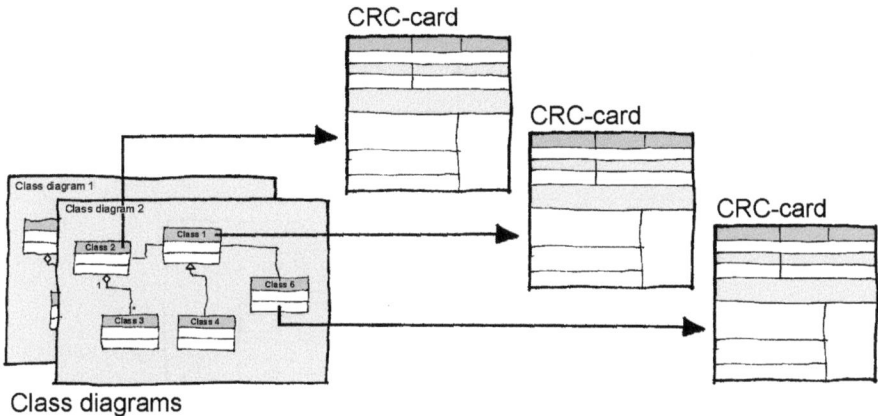

Figure 3.16. Class diagrams and CRC cards as documentation.

After implementation, the system enters the actual operational phase. When configuration systems are used, knowledge about products and possibly life-cycle systems are formalized and incorporated into an IT system. This results in a series of changes in the way in which work of specifying products and for example the manufacturing instructions is performed.

Use of configuration systems leads to a series of new tasks appearing in the organisation, while other tasks disappear. Some of the everyday tasks involving working out specifications in the form of offers and manufacturing instructions disappear. On the other hand, the sales staff have to use the configuration systems for carrying out the task of selling.

As previously mentioned, it is necessary to maintain and further develop the model that has been developed and the configuration system, in order to ensure that the configuration system preserves its validity. This means that a series of new tasks appear, which are related to maintainance and further development of the model and the corresponding configuration system. These tasks typically have to be performed by people who have the necessary knowledge of the products and the relevant life phase systems, such as production, assembly and delivery.

Maintenance and further development therefore involve a number of different people. Figure 3.17 shows an example of how the task can be organised, with one or two people (model managers) responsible for the overall model, and one or two people (programmers) responsible for updating the configuration system and the programme itself, including the user interface and integration with other systems.

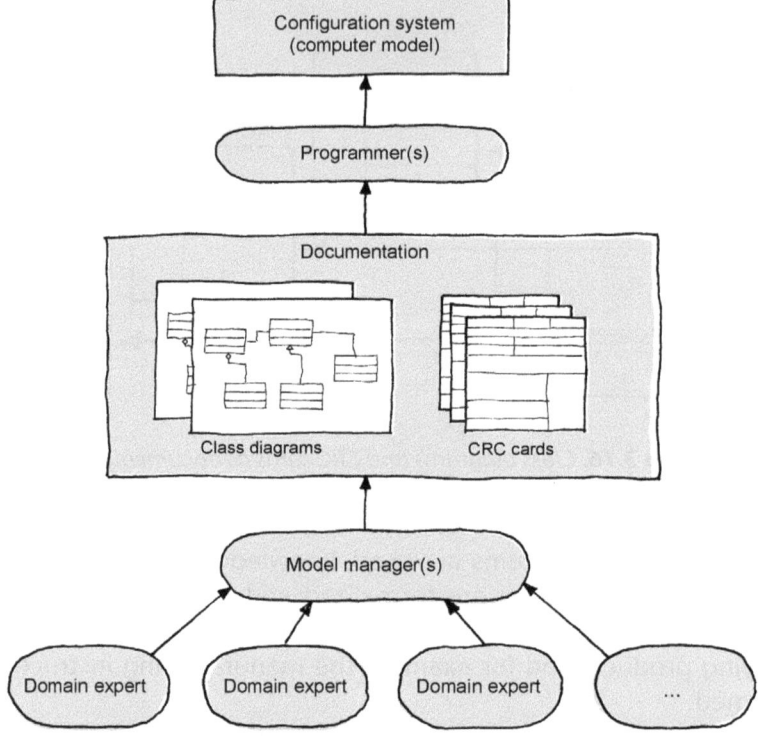

Figure 3.17. *Roles in the operation/maintenance phase.*

The people responsible for the overall models (model managers) then delegate the task of maintenance/further development to individual specialists who are each given the responsibility for maintaining that part of the model which they know about. In other words, the individual specialist is typically assigned responsibility for maintainance and further development of a series of object classes and the associated CRC cards.

In connection with development and maintenance of the models, it is essential that updating the models and the configuration system takes place at the same time. A configuration system which is not documented rapidly becomes difficult or even completely impossible to maintain and develop further.

Concluding remarks

As can be seen from what we have just presented, the phases in the procedure are not sharply separated. Work proceeds through the execution of a number of iterations, with shifts between the individual phases – analysis of the commercial requirements and analysis of the company's product range, building up a product variant master and developing an OOA model with CRC cards. At the start of the project, it may also be relevant to develop small test programmes or prototypes in order to test how a possible solution can be implemented and how it works in practice - in other words, there is an interaction between formulation of requirements and wishes and the specification of elements of the solution.

The focus of this book is in particular on the development of methods to support the analysis and development of specification processes (chapter 4), analysis of the product range (chapter 5), and modelling of the configuration system to be developed (chapter 6). These are the first three phases of the procedure up to and including the construction of the object-oriented model. The later phases in the procedure mainly follow the sequence of activities of the object-oriented project life cycle.

In connection with the choice of configuration software and programming, a more detailed introduction to knowledge representation and expert systems is presented in chapter 7, while chapter 8 describes factors relevant for the choice of configuration software. Finally, chapter 9 describes how the company F.L. Smidth has developed a configuration system to configure cement factories for the purpose of making offers.

Bibliography

[Bellin & Suchman, 1997]: Bellin David, Suchman Susan Simone; The CRC Card Book, Addison-Wesley Longman Inc., 1997.

[Bennett et al., 1999]: Bennett Simon, McRobb Steve, Farmer Ray; Object oriented systems analysis and design, University Press, Cambridge, Great Britain, 1999.

[Boehm, 1988]: Boehm W.; A Spiral Model of Software Development and Enhancement, in Thayer, R. H. (Ed.), Los Alamitos, CA, IEEE Computer Society Press, 1988.

[Booch, 1991]: Booch Grady; Object Oriented Design; Benjamin/Cummings Publishing Company, Inc., Californien, 1991.

[Booch, 1999]: Booch Grady; Rumbaugh James, Jacobson Ivar; The Unified Modeling Language User Guide, Addison-Wesley, 1999.

[Graham, 1991]: Graham Ian; Object oriented methods, Addison-Wesley Publishing Company, 1991.

[Hvam, 1999]: Hvam Lars; A procedure for building product models, Robotics and Computer-integrated Manufacturing, 1999.

[Hvam et al., 2003]: Hvam Lars, Riis Jesper and Hansen Benjamin; CRC-cards for product Modelling; Computers in Industry, January 2003 vol 50/1 pp 57 – 70.

Integrated Computer – Aided Manufacturing (ICAM), task 1 Final report (draft) Vol. 1 & 2, 1978.

[Kruchten, 2000]: Kruchten Philippe; The Rational Unified Process An Introduction Second Edition, Addison-Wesley, 2000.

Object Management Group. An organisation which amongst other things disseminates knowledge about UML via the web page www.omg.org

[Rogoll & Piller, 2004]: Rogoll Timm and Piller Frank; Product configuration from the customer's perspective: a comparison of configuration systems in the apparel industry, Conference Proceedings from the International Conference on Economic, Technical and Organisational aspects of Product Configuration Systems, DTU June 28-29th 2004.

[Warmer & Kleppe, 1999]: Warmer Jos, Kleppe Anneke, The Object Constraint Language, Addison-Wesley, 1999.

4

Development of Specification Processes

This chapter presents a proposal for how to develop a company's specification processes and define/set the scope of the configuration system(s) which are to support the individual specification processes.

As described in chapter 2, the term specification processes denotes those business processes which are related to the analysis of a customer's needs and to the definition and specification of a product which can satisfy these needs, and those business processes which specify how this product is to be produced, assembled, delivered, used, serviced and recycled or disposed of (specification of the product's life cycle properties).

The starting point for making use of product configuration should never be the technology itself, but on the contrary a recognition of the fact that the company can achieve a number of commercial advantages, such as improve quality of specifications, reduce times for making offers and carrying out orders, save resources required for making offers and order specifications, or optimisation of customized products with respect to customer preferences and production costs.

In order to concretize the possible commercial advantages and disadvantages, let us now examine the company's specification processes more closely and discuss the possibilities the use of product configuration provides for developing these business processes. Further it is described how it is possible to develop specification processes systematically and to define the configuration system(s) that will support the individual specification processes.

The process of developing specification processes

The process of developing specification processes is in general terms as follows:

Identifying and characterizing the most important specification processes

The first step is to identify and characterize the most important specification processes. The starting point is to identify the most important specifications within the company in relation, for example, to customers and suppliers, and the company's other functions, such as production, assembly and service.

Next, a survey is made of those specification processes which produce the specifications. In connection with this survey, the process' customers are identified, as well as the chacteristics of the process, with a description of important problems in the current way of working.

Requirements for the specification process

The next step involves analysing the requirements for the company's specification processes, from customers and suppliers and from other internal functions within the company. Starting from the company's business strategy and the requirements from the process' customers, the aims of each individual specification process are made clear and concrete measurements are made to show how the current specification process performs in relation to the requirements from the surroundings.

In addition to clarifying the aims of the individual specification process, it is also necessary at this stage to clarify other factors of significance for the specification process. These could for example include an evaluation of the product range's stability or the accessibility of the knowledge to be modelled. A gap analysis is performed for each individual specification process. Based on this gap analysis and the possibilities offered by product configuration for development of the individual process, it is decided which processes to focus on in the work that follows.

Design of new specification process

Step 3 involves designing the future specification processes and defining the configuration systems to support these processes.

The basis for designing the future specification process is the requirements for this process found in step 2, together with an analysis of the possibilities which product configuration offers for being able to develop the specification process. For each of the specification processes that have been chosen to focus on, one or more scenarios for the future specification process are formulated, with a description and definition of the contents of the configuration system which potentially is to support the process.

Definition of configuration system

In connection with the definition of the configuration system which is to support the specification process, the framework for modelling of product families described in chapter 2 is used to describe the general content of the configuration system.

Evaluation and choice of solution

Step 4 involves evaluating the scenarios formulated in relation to the requirements for the specification process defined in step 2, together with the expected costs and risks for realizing the individual scenarios, including an estimate of the costs associated with developing and implementing a configuration system to support the future specification process. Finally, one scenario is selected to be implemented.

Plan of action and organisation of further work

The fifth and last step involves making a plan of action and a final budget for the further work on realizing the future specification process and development and implementation of a configuration system to support the process. In addition, a plan for organising and managing the development task is worked out. Here, it is important to identify the project's milestones and the critical or most difficult phases in the execution of the project.

The individual steps in the process are not sharply separated. Analysis of specification processes and the requirements for the processes, and design of a new specification process, including a description of the configuration system(s) to be developed, will normally be carried out via a number of iterations between steps 1 and 4. In addition, it will often be useful to carry out an analysis of the company's product range, as described in chapter 5 in connection with formulation of the scenarios for the future specification process. Now let us look into the individual steps in more detail.

Step 1: Identifying and characterizing the most important specification processes

Specification processes

A business process is a structured set of activities which together give a well-defined commercial result. In this book we are concerned with the type of business processes, which with origins in the customer's product needs, produce specifications for how this product is to be designed and how the product is to be produced, transported, assembled, used and serviced (the life cycle properties of the product). We call such processes specification processes.

Specification process

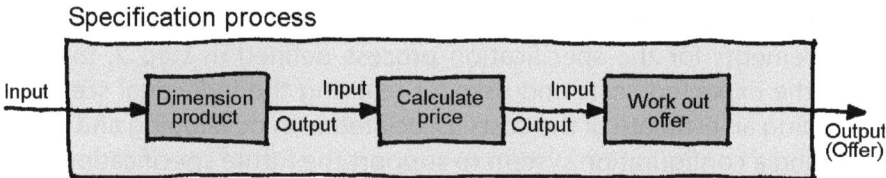

Figure 4.1. *An example of a specification process.*

A specification process can be described by its input, its output and the sub-activities performed during the process, as illustrated in figure 4.1. The individual sub-activities in the process can (like the complete specification process) be described by their input, their output and the activity performed. During the process, input, which in this case will be the customer's wishes in relation to the product, is transformed via a series of activities into output (in this case an offer).

Examples of activities in specification processes can be registration of customers' wishes, dimensioning or configuration of products, production of offers, production of drawings, production of lists of operations etc. More generally, activities are performed such as registering information (e.g. customer wishes), storing information, finding information, checking information, calculation, configuration, simulation, grouping, formulating documents etc.

A specification process can often be identified by its output – the specifications produced by the process. Typical specification processes, listed according to the specifications they produce, are as follows.

Working out offers

It is possible for this to take place in several phases, so that a rough offer (budget offer) is made first with a general specification of the product and a preliminary price. Next, one or more more detailed offers are worked out, in which the product is specified more exactly. An offer usually contains a specification of the product to be proposed to the customer. This can be in the form of a textual description, drawings, process diagrams etc. In addition, an offer will normally contain the specifications the company has received from the customer, payment and delivery conditions, and an expected delivery date and price.

Working out offers is an important part of the company's sales activity. The company's employees expend considerable resources on working out offers. In many cases, 5 or 10 offers have to be worked out in order to win an order. At the same time, the quality of the offers is critical if the company is to win the right orders.

Detailed product specification after receiving an order

After the customer has accepted the offer and placed an order, it is usually necessary in the case of complex products to make a more detailed specification of the product. Here, an offer produced previously can perhaps act as the foundation for further work on specifying the product in detail as the basis for subsequent production etc. In this context – especially in the case of more complex products – a comprehensive set of specifications can be made, for example in the form of drawings, lists of parts, process diagrams, calculation and documentation of product properties such as strength, weight, surface properties etc.

In comparison with the product specifications worked out in connection with offers, these specifications are more detailed and complete. Detailed specifications of the product form the basis for subsequent activities, such as planning, production, service or maintenance.

Specification of products' life cycle properties

Here, the specification of the product's life cycle properties is produced to define how the customized product is to be produced, assembled, transported, serviced, used and possibly disposed of or re-cycled. These specifications can be viewed as a description of the product's interaction with a life cycle system such as production, assembly, or service. Lists of operations, operation descriptions, assembly instructions and service manuals are all examples of specifications describing the product's life cycle properties.

Specification processes involve several departments

Specification processes span the company's functions, often involving staff from various departments such as sales, product development and production engineering in working out specifications for customized products in collaboration with the company's customers and suppliers.

Specification processes typically consist of activities that move frequently between departments. This also means that there are many transfers of responsibility and many occasions during the process when it is necessary to backtrack, for example in order to perform control or ask for more detailed information. The many transfers of responsibility mean that errors are often discovered late in the process, which makes it expensive and troublesome to correct them. In addition, there are often different IT systems in the different departments, which means that a lot of resources have to be used for printing out and typing in information, with the resulting possibilities for errors, inconsistency and overlap.

Specification of products involves many specialists. For example, there could be an engineer from product development who has to approve the product's configuration, or an employee from the production engineering department who has to work out production specifications. This means that an order often lies waiting in the in-tray of some specialist who is to perform a limited task.

Investigations have shown that the time needed for completing specifications for a customer order can be very long, compared to the time which is really used for working them out. The overall working time used for making a set of specifications for a customer order could for example be 4 hours, while the total period from start to finish is 4 weeks.

Value-adding and non-value-adding activities

In connection with analysing and reconstructing specification processes, one can consider which activities add value for the customer. Value-adding activities are those which directly contribute to producing a product which is in accordance with the customer's needs and expectations, and which therefore is of value to the customer. In the context of product specification, analysis of customer needs, simulation of various products, and preparation of drawings and lists of items are all examples of value-adding activities.

In contrast to these, non-value adding activities are those which do not contribute to satisfying the customer's needs, and which in principle could be removed without causing any reduction in the value of the product. Examples of non-value adding activities are:

- Checking received specifications.
- Correcting errors.
- Using time looking for information.
- Getting more detailed information from the customer at a late stage in the process.

It can in practice often be difficult to distinguish between value adding and non-value adding activities. A relatively simple way to decide whether an activity is value adding or not is to ask the question: Would the customer be willing to pay for the activity in question, if it appeared explicitly on the invoice?

The reasons for the many non-value adding activities in a specification process can for example be:

- Poor quality of the specifications produced early in the process.
- Missing information.
- Lack of an overall view of the process from start to finish.
- Inadequate IT systems.
- A confused and unstructured product assortment.

The many transfers of responsibility, reiterations, checks and the long waiting times mean that the time used for actual value adding work, in which necessary specifications are produced, often only makes up a small part of the total time which is used in connection with a specification process.

This is amongst other things due to the many non-value adding activities, which contribute both to increasing the number of transfers of responsibility and to increased use of resources. At the same time, long lead times hide many of the underlying problems and causes. Moreover, noone has an overall view of the specification process from start to finish. The customer, on the other hand, experiences the problems very directly. It takes a long time from when an enquiry is placed until an offer is received, and an even longer time from when an order has been placed until the product is received. There are often errors in the product and it is difficult to obtain a product which precisely matches the needs, which one has as a customer.

Customer focus

A specification process always has a customer. The specification process' customers can be the company's customers, who for example receive an offer with a description of the product and information about price, delivery conditions etc. But the specification process' customers can also be the functions in the company (or the company's suppliers) which use specifications to organise purchasing and to carry out production planning, production, transport, installation and after-sales service.

The starting point for development of the company's specification processes is to identify the process' customers. In relation to external customers, it is relevant to ask which customer segments or types of customer the specification process concerned is intended for. The specifications received by the company's customers can for example be brochures, offers, order confirmations, user instructions or service manuals. When the customers are identified, the requirements of the various groups of customers for the specification process and for the specifications worked out in the specification process can be investigated.

The other "customers", internally within the company or among the company's suppliers, receive the specifications necessary for the manufacture, delivery or service of the product or service concerned. A number of examples of specification processes for various "internal customers" are:

- Drawings and lists of parts are produced on the basis of input from the customer or sales department. Drawings and lists of parts are used for example in connection with production, assembly and servicing of products. In other words, the customers for the process are the subsequent processes within the company that perform production engineering, planning, purchasing, production, assembly and service. Often, several different drawings and lists of parts are produced for the individual "customers". For example, drawings and lists of parts for production are produced to describe the factors that are particularly relevant for production, while corresponding sets of drawings and lists of parts are produced which for example are targeted at assembly, installation and service.

- Production of lists of operations with the expected time consumption for each individual operation in the production process. Input for this specification process can for example be production drawings and production lists of parts. The customers for this process are production and planning.

- Production of assembly instructions. Input for this process can be drawings and lists of parts. The customers for the process are employees who assemble the company's products.

- Production of service manuals. Input for this process is first drawings and lists of parts and other specifications for the product, and secondly general directions for maintenance and servicing of the product. The customers for this process can for example be employees who perform service on the company's products. They can be employed within the company; they may be the company's external customers; or they may be employed by another company performing service and maintenance.

As can be seen from these examples, many of the subsequent processes within the company and the company's customers and suppliers are dependent on the specifications produced in the individual specification processes.

In this context it is important to understand the customers' needs, including what the customers need to use the specifications for. Here one can consider how much the specification process'- customers depend on the specifications they receive, and how much significance the subsequent processes have for the company. An example is the situation where a company chooses to move its production abroad. As a consequence, the specifications sent to the supplier abroad must be more detailed, correct and ready on time. The specification processes producing specifications for such production thus have greater significance.

Description of specification processes

The existing specification process is drawn up with the help of one of the diagram techniques described in the literature [Harvey, 2005; Holt, 2005; Khan, 2004] or one of the web pages dealing with process modelling. The individual sub-activities and their interdependencies are identified, for example by interviewing the people engaged in producing the specifications. In addition, workshops, questionnaires or role games can be used to help describe specification processes.

The aim of describing the specification process is to obtain an overall view of the most important activities in the specification process, their sequence and relationships, the people/roles with responsibility for the individual activities, the information flows through the process, and the process' input and output. In addition, any tools needed to support the process are described, such as databases, configuration systems or printed material containing information about the company's products.

In connection with describing the current process, a short characterization of the most important problems associated with the process is made – for example, that there are many transfers of responsibility or many activities connected with control or the transfer or input of information.

To illustrate the process of identification and chacterization of the most important specification processes, let us now consider the most important specification processes at Doors Inc., which we introduced in chapter 1.

Situation at Doors Inc.

At Doors Inc., customers and sales staff have been complaining for a long time that the delivery time for doors was too long; the price was too high; and the doors often had faults when they were delivered. Internally at Doors Inc. there were continual discussions about whose fault the problems were. The sales staff thought that it was the fault of the production department that the doors were always delivered so late. Production said that the orders coming from the sales staff were full of errors, and the information needed in order to produce the doors was often missing. In addition, they thought that many of the lists of parts and drawings coming from the development department were not at all suitable for production.

These attempts to place the blame had not led to any clarification of what could be done to solve the problems. To move forward, the managing director hired a consultant, who together with managers from sales, product development and production were to analyse the problems and propose possible solutions.

Everyone quickly agreed that the problems were related to the business processes lying between the customer and production. It took too long from the time the customer made an enquiry until an order could be set in production. Moreover, too many errors arose during the process.

Noone could really explain the overall process from the moment the customer made an enquiry until a complete manufacturing specification was available. In order to obtain an overall understanding of the process, the group analysed which activities were performed in connection with specification of a door.

It rapidly became clear that it was not possible to describe the process very precisely, because variations could arise from one time to the next. There was also a big difference in the process of specifying a more or less standard door and one that was more specialized. Nevertheless, people managed to agree to describe the process that covered most cases of the relatively standardized doors together with a process for those that were more specialized. However, there was no clear definition of a standard door and a special door. The process for specifying a standard door is shown in figure 1.1 in chapter 1.

The most important specifications

To get a better overall view of the specification processes, the group agreed to take as their starting point those specifications which were produced in connection with making offers and executing orders. As shown in figure 1.1, a number of different specifications have to be made before a door can be produced, transported and installed. The most important specifications produced by Doors Inc. in connection with making offers and executing orders for a customized door are as follows:

- Order requisition to be filled in by the salesman contains information about the customer's addresss, delivery address, building dimensions, and the customer's requirements with respect to the door. The customer's requirements are described by filling out the requisition with the door's dimensions, materials, colour, number and placement of doors and windows, fittings, springs, motors, controls etc.

- Offer. The offer contains a specification of the building's dimensions, the door's dimensions, materials, insulation, suspension, control etc. A sketch (not to scale) with the door's main dimensions, price, delivery address, delivery time, and a standard text with conditions of payment, valididty etc.

- Order confirmation. Refers to the offer, confirms that the order has been accepted by Doors Inc., and gives a delivery date.

- Drawing. Doors Inc. makes a sketch of the door, not to scale, indicating the door's dimensions.

- List of parts. Contains a list of the parts used in the door, together with dimensions for the individual parts.

- List of operations with times. The list of operations specifies which operations are to be performed during production, together with their sequence and time consumption.

- Production order. Starts production, and contains a specification of the parts used in the door, the door's production sequence and the delivery time.

- Installation instructions. Specifies the door's dimensions, together with the building measurements that form the basis for dimensioning the door. Also contains a series of standard descriptions of how different types of doors have to be installed, with a specification of the sequence of operations and the installation instructions.

After looking more closely at the specifications, it became clear that several of them contained some of the same information. For example, the production order contains information that also appears in the list of parts and the list of operations. The only new information is the starting and finishing time for the individual operations, together with a specification of which machine and operator are to perform the operation.

The basis for being able to complete the specifications listed above is a dimensioning of the door based on the information in the order requisition, together with a definition of the sequence of operations with a specification of the expected time consumption for each operation. When the door's dimensions have been determined, it is possible to produce a list of parts and a sketch of the door with a specification of the door's dimensions. Starting from the door's list of parts and the door's dimensions, a list of operations can be produced with calculated time consumption for each operation (time consumption for each operation can be calculated as the number of mm to be sawn * the speed of the saw in seconds/mm, or the number of holes to be bored * number of seconds per hole etc.).

When the list of parts and the list of operations have been produced, a cost price can be worked out by finding the prices of the necessary materials in Doors Inc.'s database and by multiplying the calculated operation times by a hourly price per operation.

The offer contains:

When the door's cost price is known, a letter containing an offer can be prepared for the customer. The offer contains:

- The customer's address and the delivery address.
- The price of the door.
- Specification of building measurements.
- Specification of the door with main dimensions, materials, colour, fittings, motor, springs and control etc.
- A sketch (not to scale) of the door with a specification of main dimensions and the placement of doors, windows, fittings etc.
- A standard text giving delivery conditions, period of validity of the offer etc.
- Delivery date, which as a rule is 6 weeks after date of order.

If the customer accepts the offer and places an order, he receives an order confirmation which refers to the offer, confirms that the order has been received and gives a delivery date. This delivery date is determined on the basis of the current number of orders.

Installation instructions are produced from the building dimensions, as specified in the offer, together with the list of parts and the sketch giving the door's dimensions.

Production orders are printed out by the company's planning system after the list of parts and the list of operations have been typed into the planning system.

The order office, the sales staff, product development and the production engineering department are all involved in specifying a door. In the case of specialized doors, it is often also necessary to involve the suppliers, in order to obtain information about prices and delivery times for special parts. Planning and purchasing functions are also often involved in connection with the determination of delivery times for special orders.

The customers for specification processes are partly the company's customers, who depend on the offer and the order confirmation being correct. But in addition, a series of internal functions at Doors Inc., such as sales staff, planners, purchasers, production staff, installation staff and suppliers, all depend on the specifications being correct.

The most important problems

In connection with the analysis of the most important specification processes at Doors Inc., it became clear that the problems of poor quality and delivering on time were principally due to a series of unfortunate features of the company's specification processes:

- There were many changes of responsibility, which gave many opportunities for errors to occur. And it was difficult for the next person in the chain to perceive the error and correct it before it was too late.

- The rules for how to design a door and how it should be manufactured were not clearly defined and the individual employees each had different ideas of what comprised a "good" door.

- The products were not fully designed with the intention of making it easy to create a customized variant of a door. This meant that it was often necessary to involve the development department in order to create a customized door.

- There was no clear definition of what was a standard door and what was a special door. And there were no clear general rules for how the two types of order should be dealt with.

The most important specification processes

At Doors Inc. the most important specification processes can be summarized as:

- Producing offers for standard doors.
- Producing manufacturing specifications (i.e. lists of parts, sketches with measurements, lists of operations wth times for each operation, and installation instructions) for standard doors.
- Producing offers for special doors.
- Producing manufacturing specifications (i.e. lists of parts, sketches with measurements, lists of operations with times for each operation, and installation instructions) for special doors.

These four processes involve all the doors produced by Doors Inc. for all their biggest markets, since there are no important differences with respect to dimensioning of doors for the individual markets.

Step 2: Analysis of requirements for the specification processes

After having identified and described the most important specification processes, the next step is to make clear what requirements are to be placed on these processes. These requirements come partly from the company's environment, i.e. customers, suppliers and authorities, and partly from the other functions within the company, i.e. sales, planning, purchasing, production, delivery, assembly and service, or from other specification processes, as indicated in figure 4.2.

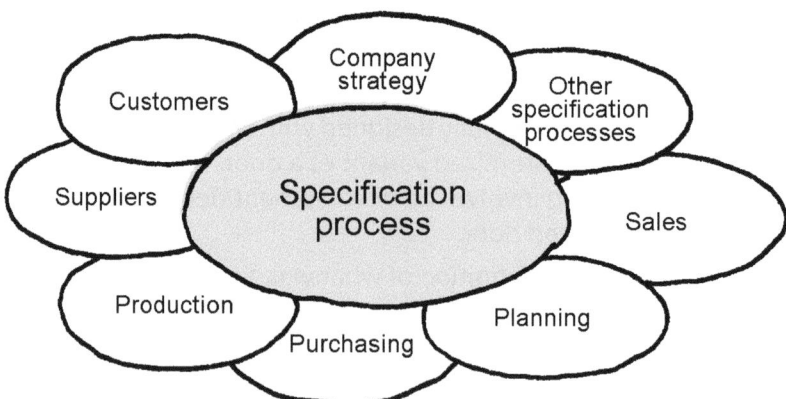

Figure 4.2. The specification process' environment.

This evaluation takes as its starting point the company's current strategy plan and the commercial targets formulated in this plan with respect to earnings, profitability, product strategy, ability to deliver, after-sales service etc. When clarifying the requirements for the specification process, some of the common tools for strategic planning, such as SWOT analysis, PEST analysis, benchmarking, etc. can be used.

It is also important to clarify requirements regarding the company's ability to offer the customers individual products.

For the specification processes identified as the most important for the company, the question is then which targets are critical if the company is to achieve its overall targets. Examples of critical targets for the process of working out offers could be that there must be at most a 5% difference between the price calculated when making the offer and the price determined by post-calculation, or that it may take 2 days at most from when a customer makes an enquiry until he receives an offer.

Five main groups of targets for a company's specification processes are:

- Lead time for producing specifications.
- On time delivery for specifications.
- Ressource consumption for producing specifications.
- Quality of specifications.
- Optimizing products/services with respect to the customers' needs or, for example, material or production costs.

In addition to these five main groups of targets, there may be a series of other factors, such as a desire to preserve a part of the knowledge possessed by the company's employees (for example in the context of a change of generations), or the wish to reduce the amount of routine work or to make knowledge available to others within the organisation. Each main group of targets is discussed in more detail below.

Lead time

Lead time refers to the interval of time from when a specification process is initiated until a finished specification is available. An example is the number of days from when a customer makes an enquiry until the customer receives an offer.

A short lead time and thus the abililty to react quickly in the market situation can often be an essential requirement for winning orders. An example is a company that produces heat exchangers to be used in larger constructions (such as cold storage in buildings or on ships), where precise delivery and rapid offers are critical parameters for winning orders. To achieve faster production of offers, and at the same time ensure registration of all the necessary product information from the customer, the company chose to develop a product configuration system for specifying products and working out offers and manufacturing specifications.

The essential condition for building up such a product configuration system was a market requirement of being able to make a binding offer from day to day. The only possibility for reducing the lead time for making offers to a single day was to develop a configuration system containing knowledge and information for working out offers.

The foundation for developing the system was thus partly a critical target of day-to-day production of offers, and partly an analysis of the lead time. This showed that this would only be possible if a product configuration system could be developed, which the sales staff could use directly at the customer's premises. An analysis of the products also showed that it would not be realistic to store knowledge and information about the product's configuration in an ordinary catalogue, partly because of the many possible combinations, and partly because of frequent changes in the product range. The requirement for a short lead time can thus in some cases be a decisive argument for supporting the specification task with a configuration system.

In this connection, it is also worth emphasizing a surprising effect of reducing the lead time. When the time is reduced, the quality of the specifications improves. This is a surprising effect, which nevertheless has been demonstrated in a long series of projects and is well documented, for example in literature about Business Process Reengineering (BPR) [Hammer & Champy, 1993; Hammer, 1996]. One of the explanations of the phenomenon is that long lead times can hide a process containing many inherent problems, such as changes of responsibility, unclear instructions or lacking documentation for the products. When the lead time is reduced, all these problems come to the surface, and it is necessary to improve matters.

On time delivery for specifications

On time delivery for specifications is defined as the number of specifications out of the total number of specifications which are completed within the agreed time span. On time delivery is normally specified as the percentage of the total specifications which are completed at the agreed time. An example is when working out offers, where the company has promised the customers that they can always expect to receive an offer within at most 3 working days. A sample of 100 random offers shows that 45 of them are delivered within 3 working days, while 55 are delivered after 4 working days or later. In this case, the percentage on time delivery for offers is 45%.

The specifications form the basis on which the specification process' customers carry out their work. For example, the production engineer can first work out process specifications when drawings and lists of parts are available, and production can first begin when all product and production specifications are available.

Thus, it is critical for the total lead time and for final compliance with the agreed time of delivery to the company's final customers that the individual specifications are finished at the agreed time. Delayed specifications also mean that subsequent processes such as production can be delayed, because employees must wait for the specifications to be finished. In other words, delayed specifications can lead to wasted time in the following processes.

Ressource consumption and frequency

The frequency of the individual specification activities, compared to the duration (use of man-hours) of the individual specification activities, reflects where the largest use of resources in the specification task lies, and where uniform tasks are executed with high frequency.

In order to be able to reveal the use of resources and find uniform tasks that are performed with high frequency, analysis can be made of the activities performed in the specification process. The analysis can be carried out as a frequency study to find out how large a part of employees' time is used on given tasks – defined in terms of the specification result or the specification method (activity).

The examples of tasks shown below are mainly defined in terms of the result of the specification task – for example that the specification produced is a list of operations or a list of parts.

Figure 4.3 shows an example of some typical tasks performed in connection with specification of products and their production. The individual subtasks are described, partly by the general activity being carried out (for example producing a list of parts or an assembly instruction) and partly by the product element involved in the activity. By a product element is meant a limited part of a product network (module) within one of a company's product families. This division is chosen because in this way the tasks are connected with a family of uniform product elements.

Activity	Product element			
	Component 1 in product family A	Component 2 in product family B	Component 1 in product family C	etc.
Produce list of parts	3 hours 200 times a year in total 600 hours			
Produce list of operations with times		1,5 hours 250 times a year in total 375 hours		
Produce assembly instruction			4 hours 30 times a year in total 120 hours	
etc.				

Figure 4.3. *Time consumption and frequency of work routines in specification processes.*

The figure shows some examples of frequency and resource consumption for activities – for example, the activity "make a list of parts for component 1 in product family A" has a frequency of 200 times a year and a duration of 3 hours per activity, for a total of 600 hours a year.

By determining how often a given task is performed and the corresponding time consumption, a picture emerges of how resources are used in the specification process, and where uniform tasks are performed very frequently.

Quality of specifications

Quality of specifications can be defined in several ways. One aspect is understandability/readability of the specifications, for example whether a customer understands the central elements in an offer he has received, or whether the production engineer understands the design drawings on which he is to base production. The basic question here is whether the specification in question is able to pass on to the receiver an unambiguous and complete description, for example, of the product's design. This aspect of a specification's quality is obviously hard to measure, both because it can be a question of subjective evaluation on the part of the receiver, and because receivers of the specification can have different backgrounds for interpreting a specification.

Another aspect of quality is the number of errors. Errors in specifications can be defined as the proportion of the specifications containing errors. Errors here are defined as those errors that, if they are not discovered, will lead for example to manufacture of a faulty product – so such errors as insignificant typos are not to be counted. An example is the number of lists of parts with errors, compared to the total numer of lists of parts produced. Another example is the number of offers in which the pre-calculated cost price differs by more than 5% from the cost price arrived at by post-calculation.

Optimizing products

Using a configuration system makes it possible to optimize products in relation to the customer's requirements, or for example in relation to production costs or maintenance/service. An example is found in F.L.Smidth, a company that manufactures equipment for cement production. F.L.Smidth has since 2000 used product configuration in connection with the task of making budget offers for cement factories. One of the most important experiences in using the configuration system is that it becomes possible to simulate various solutions and go on from there to optimize the plant with respect to operation time, emissions, or the number of F.L.Smidth components used in the plant. F.L. Smidth's use of product configuration is described in more detail in chapter 9.

As shown in figure 4.4, a series of the costs that are first actually realized later on, in connection with purchasing of materials, or use and servicing of the product, are already allocated in connection with the overall configuration of a product.

Increased insight into the consequences makes it possible to configure the product in an appropriate manner with respect to the subsequent life-cycle phases. For example, the product can be optimized with respect to production processes or material costs. Optimization can likewise be performed with respect to the product's use and maintenance.

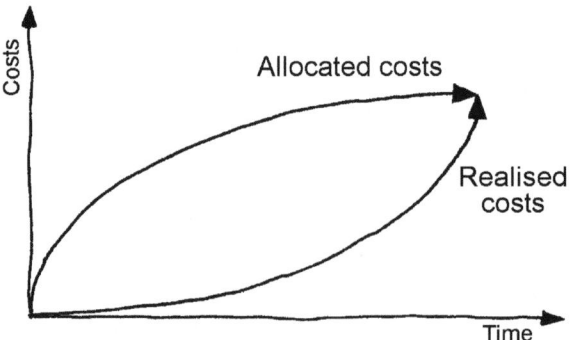

Figure 4.4. *Allocated and realized costs.*

Measuring the specification process' performance

In many companies, the targets for the individual specification processes are unknown or not formulated; it is therefore necessary to discuss the targets for the company's specification processes in relation to the company's general business strategy. This is particularly to the point when seen in the light of the opportunities which product configuration offers for developing the company's specification processes.

In this context, it can be useful to compare the targets with those of related companies and carry out benchmarking in relation to selected measuring points, such as delivery time for offers, quality of offers, or errors in specifications used in production. Benchmarking can in practice be carried out, for example, by meeting the staff of companies that are related but not direct competitors, and exchanging experiences and measurements.

In order for the targets to be able to form the basis for further work, it is necessary to make them both operational and measurable. This means that the targets must be quantified, and must be formulated in such a way that it is possible subsequently to carry out measurements within the company and to control the degree to which the targets are met. Figure 4.5 shows examples of how the overall targets for specification processes can be broken down into sub-targets:

General targets	Examples of derived sub-targets
Lead time	Lead time from customer enquiry until offer is sent to customer must be at most 2 days.
	Lead time for production of service manuals must be at most 3 weeks.
On time delivery	At least 95% of all offers must be delivered to the customer within the agreed time of delivery.
	At least 90% of all assembly instructions must be delivered within the agreed time of delivery.
Resource consumption	Employee time used on average for working out offers is to be reduced from 4 hours to 30 minutes.
	Development engineers' time consumption for support of sales staff and production is to be reduced from 30% to 10%.
Quality	There must at most be errors in 5% of the lists of parts which are sent out into production.
	When giving an offer, the price must at most differ by 10% from the price of the product/ service found by post-calculation.
Optimisation of products	In connection with progressive configuration of the product with respect to the customer's needs, the product must be optimised so that material costs are reduced by 10% with respect to the current level.
	In connection with progressive configuration of the product with respect to the customer's needs, the product must be optimised so that the use of previously designed and produced parts is increased from the current value of 30% to 60%.

Figure 4.5. *Examples of targets for the company's specification processes.*

When the targets for the most important specification processes have been formulated, the next step is to carry out a series of measurements in relation to the individual targets, so as to determine the current level of performance. When the targets and the current performance are known, the targets and performance can be summarized via a so-called gap analysis. Figure 4.6 shows an example of a gap analysis for some typical specification processes.

	Target	Current performance	Gap
Lead time	Lead time for producing offers max. 2 days	On average 8 days - large variations in lead time	6 days - to be reduced by 75%
On time delivery	95% of all offers to be delivered on time	50% of offers delivered on time	45%
Resource consumption for producing list of operations	30 minutes	4 hours	3,5 hours - to be reduced by 87,5%
Quality of lists of parts in production	95% correct	70% correct	25%
Optimisation of products	60% of parts included in product must have been produced previously	30% of included parts are produced previously	30% - number of previously produced parts to be doubled

Figure 4.6. *Gap analysis.*

Gap analysis helps us determine where the largest gaps occur between target and performance. By considering this together with the question of which targets are the most important in relation to the company's overall business strategy, we can use gap analysis as the starting point to identify which specification processes to focus on in subsequent work.

Apart from the operational targets, a number of other factors are relevant in relation to development of the company's specification processes. Some factors that can be difficult to give operational targets, but which nevertheless are important for development of the specification process, are:

- Accessibility of knowledge, including the value of maintaining knowledge within the company.
- Analysability of knowledge about products and the products' life-cycle properties – for example, whether rules for dimensioning a product variant can be expressed clearly and unambiguously.
- Mapping between functional and structural product description and continuing from there to describe relational properties (e.g. production). In other words, when a product is specified in the form of a drawing and a list of parts, for example, it is possible to make clear and unambiguous rules for how to produce a list of operations with the expected time consumption per operation.
- Stability of the existing product range and associated life-cycle systems, such as production, assembly and service.

Accessibility of knowledge

Describes who has access to knowledge about the company's products and their life-cycle properties, and how easy it is to acquire the knowledge needed. Knowledge that is incorporated into a model and structured within a configurator on a company web site is very accessible. Employees, suppliers and customers all over the world can gain access to the knowledge in question via the Internet. Knowledge that is hidden in archives or inside people's heads can be very inaccessible.

In the example of Doors Inc., the use of a configuration system will make it easier for the sales staff to gain access to knowledge about the door's design, production and price calculations. In this way, the sales person will be able to work out offers and order specifications independently.

Analysability

An important pre-condition for using product configuration is that knowledge about the company's products and their life-cycle properties can be analysed and described unambiguously.

Analysability is amongst other things connected to the complexity of the product range (number of modules, complexity of the modules, number of parameters for specification of the individual modules, requirements for variation within and between modules etc.). In the context of the product's relational properties - for example with respect to production - analysability can be described by the complexity of the overall sequence of operations in relation to one or more modules.

Specification processes are closely related to the company's product range. If the company's specification processes appear difficult to grasp and inefficient, then part of the explanation may lie in the fact that the product range is not clearly structured and precisely defined.

Mapping

Mapping denotes the relation between different representations of the product, where "representation" here refers to a representation of a product or its relation to a given system, such as production. Representations can take the form of drawings, lists of parts, operating instructions, CNC-code, lists of operations etc. Mapping thus refers to whether it is possible to map directly from one representation of the product to another.

Figure 4.7 illustrates three main groups of representations, where mapping here denotes the relation between functional product description, structural product description and production method description (e.g. lists of operations). In the example of Doors Inc., there is a simple and analysable mapping between the structural product description (drawings and lists of parts) and the description of the production method (lists of operations). This makes it possible to develop a configuration system containing rules for generating lists of operations, starting with the product's specifications (the product's structure).

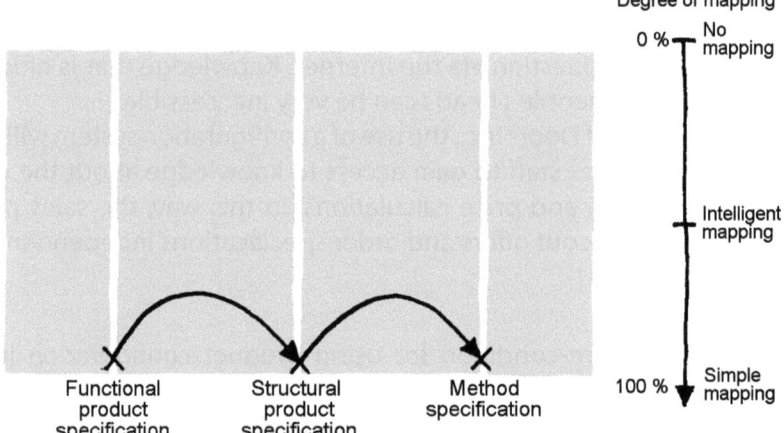

Figure 4.7. *Mapping between functional product specification, structural product specification and method specification.*

At Doors Inc., the solution space for structural product description and for method description is well defined, which is expressed as analysability. The production method can be derived directly from the structural product description by means of simple rules, i.e. there is a simple mapping. In other cases, the solution space - for example for functional and structural product descriptions – can be well defined, whereas the mapping from for example functional product description to structural product description cannot be defined.

Stability

In the context of developing a system for configuration of products, it is an essential prerequisite to be able to identify a basic product concept describing the common features of one or more families in the company's product range. By stability is meant whether it is possible to identify such a basic structure, and whether the basic product concept and production structure are stable over time.

Stability is critical for the cost effectiveness of configuration systems, since stability determines the system's lifetime, and thus the time period for depreciation of the capital invested in development of the configuration system.

When determining the basic concept for the product and its life- cycle properties (such as production methods), it is also necessary to identify the variations that with time can be expected to arise in the company's product range and production system. These can for example involve technological changes in the product or expected investments in production equipment.

In addition, other factors to be considered are:

- Knowledge sharing. When knowledge, for example about products and life-cycle systems, is formalized and incorporated into a configuration system, then this knowledge is made visible and accessible for others within the organisation, such as salesmen and employees of customers and suppliers.

- Knowledge preservation. By modelling knowledge about products and their life-cycle properties in a formal manner, the employees' knowledge becomes anchored within the company. This can for example be relevant if the company should experience a change of generations, or if there are large staff turnover.

- Reduction of routine work. The use of configuration systems will reduce the extent of routine work involved in such tasks as making offers, lists of parts etc.

- More rapid introduction of new products and easier out-phasing of old products. The use of configuration systems in sales makes it easier to implement changes in the product range, both for the customers and for the sales staff.

When the targets have been defined, the necessary measurements for defining current performance level have been carried out, and other factors of significance for the specification process have been considered, then targets and requirements for the most important specification processes can be assembled to form the basis for further work in step 3, when scenarios are made for how the individual specification processes can be developed.

In connection with this, the specification processes on which to focus in the further work are selected. Let us now analyse the targets and characteristics for the most important specification processes at Doors Inc.:

Doors Inc.: Discussion of targets

During analysis of the specification processes at Doors Inc., it became clear that there was no consensus about what the targets really were. The production department thought that the most important aspect was that the specifications to be used in production were error-free and arrived on time. The sales staff found the most important aspect to be the ability to offer the customer a door that exactly matched his needs and to give him a price and delivery time on the spot. In the orders office, they believed that it was important that the task of specifying doors should be carried out rapidly and efficiently.

An internal discussion of targets had not led to any clarification of the issue. It was therefore decided to carry out a series of customer interviews in order to discover what the customers wanted from the company. The investigation revealed that the customers were primarily interested in receiving their door rapidly and on time. Next, it was important that the door was suitable for the conditions, that the quality was satisfactory, and that Doors Inc. could deliver a door at the market price. Somewhat surprisingly, it was also discovered that only about 20% of the variants which in the course of time had been introduced in the product range, originated with the customers. 80% of the variants had arisen as a result of decisions taken in connection with purchasing and production.

Targets for Doors Inc.

Starting with this customer analysis together with an evaluation of their market standing in relation to the most important competitors, it was agreed to formulate the most important targets for Doors Inc. Figure 4.8 shows the targets, the performance delivered today, and the gap between the two.

	Target	Current performance	Gap
Delivery time: Standard doors Special doors	3 weeks 10 weeks	6-8 weeks 10 weeks or more	3-5 weeks
On time delivery: Standard doors Special doors	95 % 90 %	50 % 50 %	45 % 40 %
Quality: Doors delivered with faults (in%)	2 %	20 %	18 %
Turnover	10 % growth per year	Stagnating	10 %
Profit rate	10 %	0-4 %	6-10 %

Figure 4.8. Targets and current performance at Doors Inc.

In addition to the targets described for turnover, profit, quality and ability to deliver, Doors Inc.'s customers were making increasing demands for doors adapted to match individual requirements, for example, with respect to strength, corrosion, heat loss and design.

The stagnating turnover and profit can largely be attributed to the problems Doors Inc. has with delivering rapidly and on time, and to the considerable problems with faulty doors installed for the customers. Apart from price and design, confidence in Doors Inc.'s ability to deliver on time and without problems is the critical factor for whether the customer places an order with Doors Inc. or with a competitor.

After this, work was started to describe the targets for Doors Inc.'s specification processes. Figures 4.9 to 4.12 present the targets, performance and gaps for these processes. The targets are formulated for the production of offers for standard and special doors respectively, and for making manufacturing specifications for standard and special doors. Manufacturing specifications are the specifications necessary for Doors Inc.'s being able to plan, produce and assemble a door - in other words, a sketch with dimensions, a list of parts, a list of operations with times and installation instructions.

Measurement of current performance

In connection with the discussion of targets for Doors Inc.'s specification processes, it was noticed that there was considerable uncertainty about what the current performance was in relation to the individual measuring points. To elucidate the current performance, it was agreed to carry out a series of measurements as described below, following the principle that "one measurement is worth more than 10 opinions". The measurements were performed by taking random samples of 100 different specifications (offers, order confirmations, production orders etc.) from the previous year.

Lead time

The lead time for offers is defined as the calendar time from when Doors Inc. receives an enquiry from a customer until the customer receives an offer. The lead time for producing manufacturing specifications is defined as the calendar time from when Doors receives an order until all manufacturing specifications have been produced.

Lead time:	Target	Current performance	Gap
Producing offers:			
Standard doors	1 day	7 days	6 days
Special doors	3 days	4 weeks	17 days
Producing manufacturing specifications:			
Standard doors	1 day	4 weeks	19 days
Special doors	4 weeks	8 weeks	4 weeks

Figure 4.9. *Lead time for working out offers and manufacturing specifications.*

The delivery time for the doors is one of the most important competitive parameters for Doors Inc. Customers are willing to accept a longer delivery time for a special door, but have great difficulty accepting a delivery time of more than 3 weeks for a standard door. To achieve this target, it is necessary that the lead time for producing manufacturing specifications is reduced drastically from the current 4 weeks. Measurement of the lead time for producing manufacturing specifications is carried out by comparing the date of the order confirmation with the date for printing the production order.

On time delivery

On time delivery for offers is defined as the percentual share of the offers which have been delivered within the normal promised delivery time for offers, which is one week for a standard door and 3 weeks for special door. To measure on time delivery, the date of the customer's enquiry is compared with the date Doors Inc. sent the offer to the customer.

On time delivery for manufacturing specifications is defined as the percentage of the overall number of manufacturing specifications produced within the agreed lead time. To measure the on time delivery, the date of the order confirmation is compared with the date for printing the production order.

On time delivery is essential for Doors Inc.'s credibility in relation to its customers. The critical prerequisite for achieving the overall target for on time delivery of 95% for standard doors is that the manufacturing specifications are ready at the agreed time. Moreover, in order for Doors Inc.'s to win orders, it is critical that offers are delivered at the agreed time.

On time delivery for:	Target	Current performance	Gap
Offers:			
Standard doors	98 %	50 %	48 %
Special doors	95 %	20 %	75 %
Manufacturing specifications:			
Standard doors	95 %	40 %	55 %
Special doors	90 %	15 %	75 %

Figure 4.10. On time delivery for producing offers and manufacturing specifications.

Quality

The quality of offers is defined as the difference between the cost price calculated in advance and the cost price arrived at by post-calculation. This measurement is carried out by comparing the pre-calculated with the post-calculated price. The quality of manufacturing specifications is measured by analysing previous lists of parts, lists of operations, sketches and installation instructions, and registering major errors.

If not corrected, these errors could lead to claims for compensation – unimportant typos and the like are not counted.

Errors in specifications are one of the main reasons for low productivity and poor ability to deliver.

Quality:	Target	Current performance	Gap
Offers where cost price differs by less than 5 %:			
Standard doors	95 %	50 %	45 %
Special doors	90 %	50 %	40 %
Manufacturing specifications with errors:			
Standard doors	2 %	30 %	28 %
Special doors	2 %	30 %	28 %

Figure 4.11. Quality of offers and manufacturing specifications.

Time consumption

Time consumption is defined as the time used for producing specifications. The amount of time used for producing specifications differs greatly for the different orders. Time consumption is logged by having each employee register time consumption for a number of different tasks separately over a period of 4 weeks, such as production of the individual specifications, meetings, visiting customers, telephone conversations etc. Time consumption is then calculated as the total time consumed for producing specifications of a particular type, divided by the number of specifications of that type that have been produced during the period.

Employees who produce offers experience from time to time a marked increase in the number of enquiries from customers. During these periods, not all enquiries are answered with an offer, for the simple reason that there is no time to make more offers. Thus, there are a number of sales opportunities which are not exploited because it takes too long to produce an offer.

Time consumption:	Target	Current performance	Gap
Offers: Standard doors Special doors	0,5 hours 4 hours	2 hours 4 hours	1,5 hours 0 hours
Manufacturing specifications: Standard doors Special doors	0,5 hours 10 hours	6 hours 10 hours	5,5 hours 0 hours

Figure 4.12. Time consumption for producing offers and manufacturing specifications

Other factors

In addition to the defined targets for Doors Inc.'s specification processes for standard doors, a number of other factors are of significance for development of the future specification process. These are as follows:

Stability of the basic product concept
- The basic concept for the design and production of the doors has not changed significantly in the 30 years the company has existed, and it is expected that the doors basic construction and their production processes will be stable for the next 5-10 years. The company expects to invest in more automated production equipment in the course of the next 3-5 years.

Analysability and mapping
- In connection with the analysis of Doors Inc.'s specification process, an attempt has been made to work out a set of rules for a limited part of a door family for how to design a door variant and specify a list of operations with calculated time consumption. The attempt demonstrated that it is possible to make a series of relatively simple,

well-defined rules for dimensioning doors and their fittings, springs, motors, controls etc. It also turned out that once a door's list of parts had been specified, a list of operations could be produced on the basis of a total list of operations, by selecting or omiting the relevant ones, based on information about the door's list of parts. Operation times could be calculated for all cases from the door's list of parts and the main dimensions by using simple formulas.

To continue the process, the decision was made to focus on making offers and manufacturing specifications for standard doors. The reason for this was partly that standard doors make up 80-90% of the turnover, and partly recognition of the fact that it would not be possible to develop a configuration system that could deal with all possible variants of a door.

As a foundation for the further work, it was decided to initiate a process for defining a standard door and a special door. In order to obtain an overall view of the product range, the product range for standard doors was analysed and described by using a product variant master, as described in chapter 5.

Step 3: Designing the future specification process. Definition of configuration system

When the most important specification processes have been identified, the next step is to design the new specification processes and to define the configuration system(s), which are to support the individual specification processes.

When designing the coming specification process, it can be useful to pose a series of questions about how to work in the future:

- Which targets should be met by changing the specification process?
- Where do the greatest potential opportunities lie for developing the specification process?
- What is the nature of the specification work to be performed?
- Are there methods to be used in the specification process stable over time?
- What degree of freedom and flexibility must be inherent in the specification process and the configuration system?
- To what degree of detail must one work in the specification process?

- What types of product knowledge (cf. the framework for modelling product families described in chapter 2) should be entered into a configuration system?
- How is input from the customers secured? When a configuration system is introduced for executing offers, a risk exists for losing customer contact and thus valuable input from customers. On the other hand, a configuration system can contribute to a more systematic collection of customer input. Also, regular customer visits or other methods for systematically ensuring a dialogue with customers can be considered.
- How is continued innovation of the products and specification processes ensured by using a configuration system?

In connection with the task of analysis, many ideas will be presented about how to use product configuration for the job of producing specifications. In order to impose some structure on the discussion, it can be useful to summarize the ideas in the form of scenarios for how the specification processes can be developed.

Proposals for the future specification process are drawn up by using the same diagrammatic techniques as those used to describe the existing specification process. The process' input and output are defined. Input can be described as a list of information that is necessary to be able to make the specifications that comprise the output of the process. The process' output is defined by describing the specifications (such as offers, lists of parts for production or lists of operations for production with calculated time consumption per operation). In this connection it is important to work out a complete description of the specifications that comprise the process' output, with a description of which information is to be generated specifically for the individual order.

Next, it is described how the specifications are to be produced, which of them are to be generated by a configuration system, and who is responsible for production of the specifications. As part of this task the configuration system's overall scope is defined (system definition), which includes:

- The purpose of implementing the configuration system.
- A description of the specification processes which the configuration system is to support.
- The system's input and output.
- The system's user interface.
- Integration with other systems.

- The system's functionality; functions are assigned "need to have " and "nice to have" priorities.
- The knowledge to be incorporated into the configuration system (cf. framework in chapter 2).

In addition, a series of further requirements for the configuration system can be formulated, such as response times, number of users, language etc., as described in chapter 8.

To define the tasks to be performed by the configuration system, the starting point can be the system's expected input and output. The system's output can typically be defined by the specifications which the configuration system is to produce, for example offers or lists of parts. The system's input can be described by defining the screen images for typing in product information, and by defining the integration with other IT systems that might deliver input to the configuration system.

The next step is to define which calculations are to be performed by the configuration system – i.e. how the configuration system can deliver the expected output. Here, it can often be a good idea to produce a prototype of the configuration system, in order to test whether the system can handle the task technically. It is also necessary to define how the configuration system is to communicate with other systems, such as the company's ERP system, CRM system, or programmes for dimensioning products.

For describing the configuration system, the framework for modelling product families can also be used (described in chapter 2), also to define which product families and/or product modules should be included in the models and what degree of detail should be used.

Doors Inc.: Designing a new specification process

To achieve the bold targets set for the specification processes at Doors Inc., it would be necessary to perform a drastic reorganisation of the way Doors Inc. specifies products. Knowledge of the doors and their production procedure must be formalized. In addition, it will be necessary to investigate the possibility for reducing the number of changes of responsibility by using IT to support the process.

In connection with the analysis of the requirements for Doors' specification processes, many good ideas emerged for how the specification processes could be improved. In order to structure the discussion, the suggestions were collected into three different scenarios for the future specification processes.

Scenario 1

The first scenario was in general terms based on preserving the current work procedures and instead improving the order requisition and other documents. At the same time, the sales staff and those who handled orders were to be trained in how to use the product assortment and how the various documents should be filled in. A clear separation would also be made between standard and special doors, and a formalized work procedure should be worked out for the two types of orders.

Scenario 2

The next scenario was more daring. The idea was to develop a configuration system which, from the starting point of input from the sales staff, could dimension a door and print out a list of parts and a sketch with dimensions. In addition, the system should be able to produce and print out a list of operations with calculated time consumption, together with an offer and installation instructions. Figure 4.13 shows the configuration system's input and output.

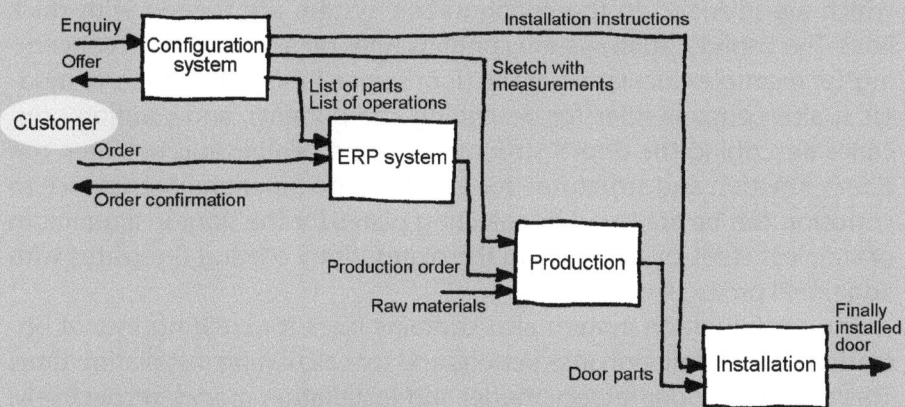

Figure 4.13. *The configuration system creates the specifications, which have to be worked out for each customer order.*

The configuration system's input is entered in connection with the production of offers and the setting up of an order, and includes the door's main dimensions, choice of materials etc. The system's input is typed in by the salesman or those dealing with orders, in connection with a customer enquiry, using the customer's input.

The configuration system's output comprises an offer, a list of parts, a list of operations with calculated time consumption per operation, a not-to-scale sketch of the door with the door's main dimensions, and a set of installation instructions.

The configuration system contains a structural description of the doors, including a description of functional properties such as heat loss and corrosion resistance. In addition, the configuration system contains knowledge about the doors' production and installation.

With respect to making sketches with dimensions, about 500 drawings of doors are made, after which the configuration system selects the drawing which lies closest and adds the door's dimensions to the drawing. With respect to calculation of the cost price for the materials used, once or twice a month a copy is made of the stock database with prices from Doors Inc.'s ERP system, and this database is used in connection with the configuration system to calculate the cost of the door's materials.

Lists of parts and lists of operations with times for the individual orders must be transferred from the configuration system to the ERP system each time a customer order is created.

Figure 4.14 shows the configuration system's overall content in terms of the framework described in chapter 2. Those parts of the framework which are included in the configuration system are framed with thick lines. The configuration system contains rules on a generic level concerning, for example calculation of the door's price (under the property model). It also contains rules for describing components, and solution principles describing the door's structure and the relationship between the door's function and structure - for example, the function of resistance to corrosion can be achieved by selecting plates for the door in aluminium or stainless steel, by galvanizing the plates, or by coating the plates with additional protective lacquer.

The configuration system also contains rules for creating lists of operations and installation instructions and for calculating installation time. This is shown under factory model and installation model, respectively, on the generic level.

On the instance level, the configuration system must use a database with about 500 door drawings, together with a database with up-to-date prices for the materials used in the door. This is shown on the instance level under the part model and property model, respectively.

Scenario 3

The third scenario was in many ways reminiscent of scenario 2 but was more ambitious with respect to creating drawings and integration into

	Property models (Derived properties)		Product structure model		Models of the product's meeting with life cycle systems				
	Internal and external properties	Functional properties	Solution principles	Part model	Factory model	Process model	Transport model	Installation model	Other life cycle models
	Describes consequences of meeting between product and life cycle systems	Describes product's function	Describes product's function-bearing units	Describes product's components	Overall description of production equipment, layout etc.	Detailed description of individual manufacturing processes and production equipment	Describes transport of product	Describes installation of product	Can for example be service or disposal / recycling
Generic level	Rules for calculating price and weight	Rules for calculating corrosion properties / Rules for calculating insulation / heat loss	Solutions in principle for corrosion protection / Solutions in principle for insulation	Rules for selection and dimensioning of the door's components	Rules for selection of operation site and rules for calculating time consumption	Rules for describing the individual production processes	Rules for selecting form of transport and calculating transport price	Rules for formulating installation instructions and calculating installation time	
Instance level	Tables of prices for parts to be used. Tables of weights of parts to be used.	Functional description	Description / definition of solutions in principle	Drawing, list of parts etc.	List of operations, production layout, description of production equipment etc.	Process description, description of tools, CNC code etc.	Transport price, description of packaging, transport documents, etc.	Installation instructions, list of installation equipment, installation time etc.	

Figure 4.14. The overall content of the configuration system at Doors Inc.

the ERP system. When configuring the door in scenario 3, a 3D drawing of the door can be automatically created, drawn to scale and containing all the door's dimensions and all the parts used. In relation to the ERP system, a dynamic integration between the configuration system and the ERP system can be made, so that the configuration system operates directly on the ERP system's stock database.

Step 4: Evaluation and choice of solution

To evaluate the individual scenarios, an evaluation is made for each scenario of the commercial advantages and disadvantages to be achieved by changing the specification process, including advantages and disadvantages associated with implementing a configuration system. In addition, a critical evaluation of the most important risks involved in implementing the individual scenarios is performed.

Evaluating the individual scenarios can start with the targets set up for the specification process. The individual scenarios are evaluated with respect to how much they contribute to achieving the targets formulated for the specification process. In this connection, it would also be relevant to compare the future specification process' performance with other companies or branches of industry by benchmarking.

An evaluation is then performed of the costs incurred by the individual scenarios. These are partly the project costs associated with changing the specification process and possibly developing a configuration system to support the process, and partly the running costs involved in operating the specification process.

Project costs can be divided up into hourly costs for the company's employees in connection with the execution of the project, consultant costs and software costs. The costs for running the specification process involve the time consumption for making specifications, time consumption for maintenance and further development of the configuration system (divided into external consultant hours and company hours), together with expenditure for software licences.

To be able to evaluate the proposed scenarios, it is also important to evaluate the greatest risks associated with each scenario. In the context of the use of product configuration, a scenario's risks can be divided into risks associated with developing a configuration system, risks associated with deploying and using a configuration system, and finally risks associated with maintenance and further development of a configuration system.

Some examples of risks associated with developing a configuration system are as follows:

- It can be difficult to gain access to the knowledge necessary to develop a product model, for example because the staff does not possess this knowledge or will not share the knowledge with others.
- Nobody feels a sense of ownership for the configuration system. The project is not sufficiently deeply rooted in the minds of the right people in the company.
- Parts of the product model can be difficult or impossible to implement in a configuration system
- The configuration system becomes too large and complicated, because a suitable effort to limit and focus the system has not been undertaken.

In connection with deployment of the configuration system, examples of risks could be as follows:

- The users of the configuration system have not received the necessary training in the products or the system and are therefore uncertain about how to use it.
- The configuration system has not been tested and corrected before the users are given access to the system. This can lead to refusal to use the system, since they receive the impression that it is too insecure.
- The users lack motivation for using the system.

Maintenance, error correction and updating/further developing the configuration system are of vital importance for ensuring its successful use. In this connection, the risks can be as follows:

- Production of suitable documentation and structure for the configuration system has been neglected. This can make it impossible in the long run to maintain or further develop the configuration system.
- Staff - for example in the company's development department - lack the necessary commitment to contribute to maintaining and further developing the configuration system.
- Maintenance and further development of the configuration system are neglected. As a result, the system's users lose confidence that the configuration system contains the latest updated knowledge on the company's products, and therefore do not wish to use the configuration system.

These are just examples of risks. Before initiating a configuration project, it is important to identify the greatest risks in the project and take the necessary steps for countering and minimizing them. A configuration project's risks can typically be reduced by following a systematic procedure, by using professional project management, and finally, by using principles for change management [Kotter, 1996] in order to handle the organisational changes which take place as a result of developing and implementing a configuration system.

Doors Inc.: Evaluation of scenarios

After having created the three scenarios for configuration of standard doors at Doors Inc., an attempt was made to evaluate possible commercial gains in relation to the objectives of step 2, together with the costs and important risks in the three scenarios as summarised in figure 4.15.

	Target	Scenario 1	Scenario 2	Scenario 3
Lead time for: - Offers - Manufacturing specifications	1 day 1 day	5 days 5 days	1 day 1 day	1 day 1 day
On time delivery for: - Offers - Manufacturing specifications	98 % 95 %	70-80 % 70-80 %	99 % 99 %	99 % 99 %

Quality:
- Accuracy for offers
- % errors in
  manufacturing
  specifications
	95 %
	2 %
		60-70 %
		15-20 %
			95 %
			2 %
				95 %
				2 %

Productivity: - Offers - Manufacturing specifications	0.5 0.5	1,5 hours 5 hours	0.5 0	0.5 0

Figure 4.15. *Evaluation of the expected effects of following scenarios 1-3.*

Expected effects in relation to objectives

Implementing the first scenario would only have modest effect. There is also a considerable risk that the staff, after having carried out a project involving formalization of business processes, would fall back into their old bad habits. Thus, it would be difficult to achieve a lasting improvement.

Scenarios 2 and 3 both have a marked effect in relation to the proposed targets. There are no obvious large differences between the two scenarios. A 3D scale drawing of the door provides a better basis for the staff in production and for the installation workers. A 3D drawing could also be used directly as a basis for programming the automatic production equipment in which Doors is expected to invest in the course of the coming 3-5 years.

Evaluation of costs for the three scenarios

Figure 4.16 presents a rough estimate of the expected project costs and costs per year for operating the specification process for each of the three scenarios. The project costs include both the hours needed for development work (divided into Doors Inc.'s own employees' time and external consultant services), and the expected costs of investment in software.

Running costs include the employee hours used for each of the three scenarios for making specifications. For scenarios 2 and 3, expected expenditure for software licenses and costs for maintenance and further development of the configuration system are also included.

During evaluation of the time consumption necessary for running the specification process for the three scenarios, it was estimated that in the first scenario about 10 people would be needed to produce offers and manufacturing specifications. In the second and third scenarios, it is expected that this could be reduced to 3 people, primarily to produce offers and manufacturing specifications for special doors. It was also assumed that in scenarios 2 and 3, one person would be employed full time to run and maintain the configuration system.

For the calculation of time-related costs, Doors' own employees' time has been rated at 80 euro/hour, while consultant hours are rated at 150 euro/hour. For calculating software costs for scenarios 2 and 3, the costs for software licenses have been set to 20% of the initial price of the software. Investment in software is set to 70,000 euro in scenario 2 and 100,000 euro in scenario 3. The software costs in scenario 3 are subject to some uncertainty.

When evaluating the three scenarios, it should be noted that scenario

1, which is close to a continuation of the current way of operating, is by far the most expensive of the three scenarios in terms of running costs.

	Project costs		Running costs	
	Hours	EUR	Hours	EUR
Scenario 1:				
- Own hours	400	32,000	15,000	1,200,000
- Consultant hours	100	15,000	0	0
- Software		0		0
- Total		47,000		1,200,000
Scenario 2:				
- Own hours	3,000	240,000	6,000	480,000
- Consultant hours	500	75,000	200	30,000
- Software		70,000		14,000
- Total		385,000		524,000
Scenario 3:				
- Own hours	6,000	480,000	6,000	480,000
- Consultant hours	1,500	225,000	400	60,000
- Software		100,000		20,000
- Total		805,000		560,000

Figure 4.16. Estimates of project and running costs for the 3 scenarios.

Evaluation of risks for the three scenarios

The most important risks were evaluated for each of the three scenarios. The evaluation of risks included both an evaluation of risks related to the execution of the development project for the specification process and the possible development of a configuration system, and risks related to the subsequent implementation, operation and maintenance of the future specification process and the associated configuration system.

Risks in scenario 1

The greatest risk in the first scenario is that the employees consider the project as purely a rationalization project, where the focus is on having the individual employee produce more in the same amount of time. In connection with the operation of the specification process in scenario 1, there is a considerable risk that employees would fall back into their old work routines. Then, it would be difficult to achieve a lasting improvement in the process' performance.

Risks in scenario 2

In the second scenario, uncertainty exists about to what extent it will be at all possible to develop a configuration system that can dimension a door. To counter this uncertainty, the decision was made at an early stage in the project to programme prototypes of the system, in order to check the critical parts of the configuration system. It was also decided to contact a consultant who had previously developed and implemented configuration systems to work on the project.

With respect to implementation and operation in scenario 2, there is also a risk that the sales staff and employees in the order office at Doors Inc., the primary users of the system, would not start using the system. The reasons for a possible rejection of the configuration system can be both that the system is not able to deliver the necessary support for the task, and that the employees are uncertain with respect to how they shall use the system and how their job situation will be affected.

To counter this uncertainty, the users of the system are to be involved at an early stage of the process and trained to use the system. The company also considers introducing a bonus system to reward the users of the system.

Risks in scenario 3

In the third scenario, the technical risks are markedly greater, especially with respect to integration with the ERP system and production of a 3D drawing. The estimated time consumption is associated with great uncertainty, and there is a real risk that parts of the project can simply not be realized technically.

With respect to operation of the system, more or less the same risks exist as in scenario 2. Figure 4.17 summarizes the main risks in the three scenarios.

Evaluation of advantages, costs and risks for the three scenarios indicates that scenario 2 will give the best results for Doors Inc.

	Execution of project	Operation of specification process and configuration system
Scenario 1	Risk that project is considered as "pure rationalisation"	Risk of falling back into old routines
Scenario 2	Small to moderate risk that parts of configuration system cannot be realised	Small to moderate risk that sales staff and order processors will not use system
Scenario 3	Moderate to large risk that parts of configuration system cannot be realised	Small to moderate risk that sales staff and order processors will not use system

Figure 4.17. The most important risks in the three scenarios.

Step 5: Plan of action and organisation of further work

When the scenario to be realized has been chosen, a plan of action for the task of development is worked out. This plan contains a description of the tasks to be performed, with specification of the most important tion of the resources to be used in the project, and the way in which the work is to be organised. In addition, the project's internal and external stakeholders and their expectations for the project are identified.

The first step is to describe the tasks to be performed. The project's tasks are apparent from the procedure for developing configuration systems described in chapter 3. Some of the typical tasks that arise in connection with the development of specification processes and the construction of configuration systems are as follows:

- Analysis and modelling of product range.
- Determination of work procedures in the future specification process.
- Definition of the configuration system's input/output, i.e. the user interface, and the documents to be produced by the configuration system.
- Specification of integration with other systems in the company.
- Programming of product configuration system and the system's user interface.

- Programming of integration with other systems within the company.
- Test of configuration system, user interface and integration modules.
- Training of the system's users in the configuration system, in the company's product range and in the specification processes.
- Training of the domain experts who are to develop and maintain the configuration system.
- Management of the project for developing specification processes and configuration systems.
- Follow-up and further development of specification processes and configuration systems.

For each task, it is determined who is responsible for performing the work, who are the other employees involved and their expected time consumption, and the starting and finishing times for the task. In connection with manning the project, distribution of the various roles in the project (described in chapter 3) should be considered. With respect to the project's organisation, it is important to ensure the necessary support by management and the project staff. In this context, it can be useful to analyse the project's stakeholders, their needs and expectations with respect to developing and using a configuration system, as well as how they can be influenced or drawn into the project. In addition, a budget is worked out for the project.

In the context of working out a plan of action for the project and the subsequent management of the project, it is important to focus on the parts of the project where a risk can be anticipated that the project will not achieve its defined targets. The question must be asked: "What is the most difficult obstacle for realizing the project?" It could for example be parts of the configuration system, where there is doubt about whether from a technical point of view it is possible to programme the configuration system; or it could be in relation to the users' acceptance and use of the configuration system.

If it is the first time the company is developing a configuration system, a considerable degree of learning will be necessary during the development process - for example learning in relation to the configuration technology, learning to know the company's products, learning how to use a systematic procedure, or learning about the organisational changes that take place when a product configuration system is used.

In this connection, it can be a good idea to carry out the project in the form of a number of iterations, as shown in figure 3.7. In this way, it will be possible to collect the experiences learned during the process, and possibly correct the project's objectives.

In connection with operation and maintenance of the changed specification process and the associated configuration system, it can be a good idea to perform measurements of the process' performance, as described under step 2. In this way, the commercial advantages achieved are made visible, and at the same time, this provides the possibility of intervening at an early stage if the expected targets are not attained.

Parts of the action plan for further work at Doors Inc. are as follows:

Doors Inc.: Action plan for performing scenario 2

In connection with the realization of scenario 2, the procedure described in chapter 3 is followed. To reduce the project's risks as much as possible, programming and testing the parts of the configuration system considered to be the most critical for the project are gradually performed. In addition, the system's coming users – sales staff and people handling orders – are involved in the task from the very start of the project, and the necessary resources are set aside for training the system's users.

Activity	Resource people	Person responsible	Expected time con- sumption	Start	End
Analysis and modelling of product range	Performed by project leader and consultant, and about 10 staff members fram sales, product development, production, installation and order processing.	NN	20 man weeks	1.3	1.5
Programming of configuration system	Performed by programmer in collaboration with project leader and consultant.	NN	10 man weeks	1.4	1.6
Test of configuration system	Project leader and programmer, together with members of staff from sales, product development, production, installation and order processing.	NN	5 man weeks	1.5	1.7
Definition of work processes in the future specification process	Project leader and consultant in collaboration with sales staff and order processors.	NN	10 man weeks	1.3	1.5
...					

Figure 4.18. Plan of action and expected use of resources for scenario 2.

Figure 4.18 shows some of the activities in the action plan for scenario 2. The figure also shows which employees are to perform the individual activities, the expected time consumption, the starting and finishing times, and who is responsible for the activity concerned. As indicated in figure 3.7, product analysis, modelling, programming and particularly testing are performed as a number of iterations.

The project is led by one employee from Doors Inc.'s development department. The project leader is responsible to the director at Doors Inc., who is the project's sponsor. The project group contains employees (domain experts) from Doors Inc.'s development department, sales department, order processing department, production and installation, each of which contributes with knowledge about customers' requirements for the doors, the doors' design, production and installation. In addition, sales staff and staff from order processing are involved in connection with definition of the system's user interface. From outside Doors Inc., an external consultant with experience in developing configuration systems also contributes.

The organisation of the development project is shown in figure 4.19. In relation to the roles described in chapter 3, it must be noted that the project leader, apart from day-to-day project management, also takes on the roles of innovation manager, model manager and process manager. Thus, it is for example the project leader's responsibility to ensure that the dialogue with the configuration system's users takes place, and together with the configuration system's users to create the future specification process.

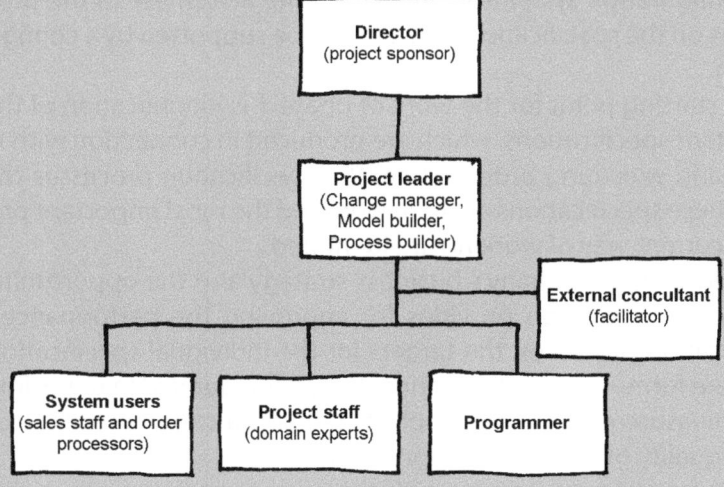

Figure 4.19. *Organisation of development project.*

At the end of the development project, the specification process and the configuration system enter the operational phase. In connection with operation of the system and the future specification process, the following tasks must be emphasized:

- Measurement of performance in relation to the defined targets for the specification process.
- Input from Doors Inc.'s development department about updating the product model and the configuration system.
- Maintenance of the product model and configuration system.
- Running dialogue with the configuration system's users and continual adaptation of the configuration system and the system's user interface.

In the operational phase, an employee from the development department is responsible for maintaining the product model. In addition, an employee is attached to the project with the day-to-day responsibility for updating the configuration system. Employees from product development, sales, production, assembly and the order processing office are also attached to maintain the configuration system via a configuration team that meets about once a month. The operational phase is organised as shown in figure 3.17.

Summary of phase 1 in the procedure

In this chapter, the content of phase 1 of the procedure for developing configuration systems is presented. The first phase of the procedure focuses on the specification processes to be supported by a configuration system.

The starting point for the work of phase 1 is identification of the most important specifications, which are produced in connection with making offers and executing orders. Then, the specification processes that produce these specifications are reviewed, and the most important problems in the current way of working are identified.

Based on the company's business strategy and the opportunities that product configuration provides for improving the performance of the specification processes, the targets for the individual specification processes are formulated. In this connection, it is a good idea to perform concrete measurements, for example of the specification process' lead time or the quality of the specifications produced.

By comparing the targets with the current performance, a diagnosis can be made of where an effort is recquired in order to improve the specification processes. On this basis, the future specification process can be constructed and the configuration system that will support the process can be defined. Here, it can often be a good idea to carry out an analysis of the company's product range (described in the next chapter), and if necessary work out one or more prototypes of a configuration system, in order to investigate the technical possibilities. Then, the future specification process and the configuration system that will support the specification process are summarized in one or more scenarios.

The individual scenarios are evaluated in relation to the defined objectives, together with the costs and the greatest risks involved in realizing the scenario. Then, the scenario to be implemented is selected, and an action plan for further work is formulated, including the most important milestones, a organisational plan, and a budget for the development task.

Bibliography

Business Process Management Group 2005: In Search Of Bpm Excellence: Straight From The Thought Leaders; Meghan Kiffer Press, Tampa, Florida, USA.

[Hammer, 1990]: Hammer Michael: Re-engineering work: Don't automate, obliterate, Harvard Business Review, July-August, 1990.

[Hammer & Champy, 1993]: Hammer Michael, Champy James: Reengineering the Corporation - a manifesto for a business revolution, Harper Collins Publishers, 1993.

[Hammer, 1996]: Hammer Michael: Beyond Reengineering, Harper Collins Publishers, 1996.

[Harvey, 2005]: Harvey M.: Essential business modelling, O´Reily, Farnham, USA, 2005.

[Holt, 2005]: Holt J.: A pragmatic Guide to Business Process Modelling, British Computer Society (BCS), Swindon, UK, 2005.

[Khan, 2004]: Khan R.N.: Business Process Management: A Practical Guide, Meghan-Kiffer Press, Tampa, Florida, USA, 2004.

[Kotter, 1996]: Kotter John P.: Leading change, Harvard Business School Press, 1996.

Web sites about process modelling:

BPMN http://www.bpmi.org

UML activity diagrams http://www.omg.org

IDEF0: http://www.idef.com/idef0.html

5

Analysis of the Product Range

One of the big challenges in building up a configuration system is to create an overall view of the range of products to be incorporated into the configuration system. A product range is normally created through developing a series of concrete products to suit different customer needs. In this way, a product range is created gradually over a number of years, as a result of input from different customers.

Thus, if you ask for a description of the company's product range, you will often not receive an answer. At the very least, you will have to ask more than one person in order to get an overall view of the product range. Different attitudes may exist about the company's product range. If the same customer were to approach two different engineers within the company to design a customized product, the customer would in many cases end up with two quite different products.

A configuration system is an IT system which can configure a customized product on the basis of a series of inputs about which requirements the customer has for the product. The configuration system thus contains a description of the company's product range and rules for how a customer-specific product variant can be designed. As indicated above, these rules are often not clearly formulated, and there can be many different ways of understanding them.

It is therefore necessary to carry out a process within the company in which all relevant stakeholders (sales staff, product developers, production staff, purchasers etc.) contribute to producing an overall picture of the complete product range and agree on which product variants the company wants to offer via the configuration system.

In the following sections, we present a methodology to support the task of defining the product range to be incorporated into the configuration system. As our starting point, we discuss the various points of view that can be used when examining a product range on the basis of system theory. Then, we present the technique of modelling a product range.

Experience from a long series of configuration projects has shown that the structure of the knowledge described in the configuration system is of great significance for a large number of activities, such as maintenance, performance of the configuration system etc. Some of the central questions in connection with the product analysis are:

- Is the product range ready to be dealt with in a configuration system, i.e. what degree of modularization exists in the product range?
- How are the limits set for that part of the product range that is included in the configuration system?
- How do we get a general overview of the knowledge to be dealt with? Knowledge will often be distributed over a large number of IT systems, for example CAD, ERP, PDM, SCM and CRM, as well as in the heads of a number of different product specialists.
- How are the product model and the configuration system to be structured, so the system will be easy to maintain and offer an optimal division of work between the configuration system and users?
- How do we ensure the quality of the knowledge to be modelled in the configuration system? There are often conflicting points of view within the company concerning rules, the degree of detail etc.

This chapter is concerned with the above questions and presents an operational tool - a so-called product variant master - for modelling and visualising a product range. In simplified terms, it can be said that a product variant master contains a description of the product range and the associated knowledge all described on one large piece of paper.

In connection with configuration projects within companies, we often meet the attitude: "We know our product range well and also what is to be dealt with in the configuration system. Why don't we just get going with programming it in the configuration system?" Amongst the most important arguments for carrying out a thorough product analysis are:

- Knowledge to be modelled in a configuration system is normally spread out among several experts, for example, from sales, development and production. It is vital that the configuration system reflects relevant stakeholders' interests. If for example it is a sales configuration system, there will be many stakeholders other than the sales staff who play a role in running the configuration system. If modelling takes place directly into the configuration system, it will be difficult for non-IT experts to see whether the knowledge which is included is also the right knowledge.

- It is very important that the right level of detail for describing product elements and rules is identified, based on the need for which the configuration system is to provide support. There are many examples of development departments who have built up configuration systems for sales purposes. This can mean that the level of detail is too high, and that the configuration system does not support the actual need.

- It is vital that a knowledge structure which suits the configuration system's stakeholders is identified. When the configuration system is once established, it can be difficult to change the knowledge structure, which therefore remains more or less unchanged throughout the configuration system's lifetime. It is therefore vital to identify the areas in which knowledge is stable over time, and correspondingly also those areas where knowledge frequently changes. Areas with frequent changes can be components, whereas calculation principles can be more static over time.

- When the structure of knowledge is identified, decisions are also made about the use of the configuration system. For example, many of the rules for configuration of products determine to a considerable extent the dialogue which is created between the configuration system and the use of the system. It is therefore essential that the project group can see the rules that are included, and that the consequences of using the rules in the configuration system are discussed with the stakeholders.

In other words, as in many other projects, the degree of success or failure is determined in the early phases of the project. It is therefore important that a thorough product analysis is carried out before the modelling of knowledge in the configuration system is started. The effort spend in the product analysis is "repaid" many times in connection with the programming of the configuration system and later during operation and maintenance.

The product programme's readiness for configuration

A question often arising during a company's considerations about product configuration is to what extent the product programme is ready to be incorporated into a configuration system. There seem to be two approaches, if the product programme is too complex for the company and perhaps also the customers to understand.

One approach is to start a modularization and standardization project before starting a configuration project, so that basically a "clean up" is performed in the product programme and the associated IT systems.

The other approach is, for example, in a sales configuration system, to consider which variants are to be offered to the customers. After this, it is "market mechanisms" that decide which variants of the products are needed. We have seen companies achieve success with both approaches, and it is difficult to give one preference over the other.

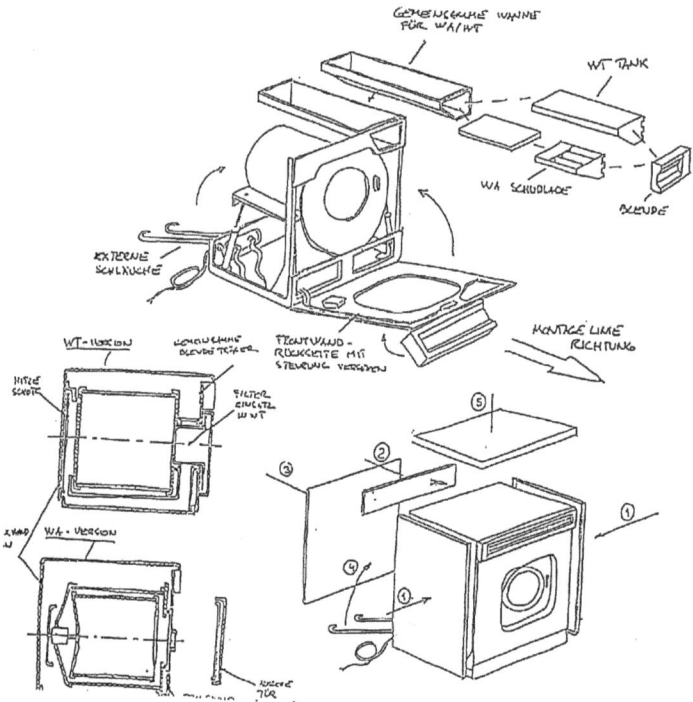

Figure 5.1. Modularization of a washing machine and drier [Fabricius, 1994].

Experience from previous configuration projects indicates that, in order to achieve marked advantages via configuration, it is a necessary pre-requisite that the company has complete control of the product programme, IT systems and the configuration processes that are to manage the products. It lies outside the limits of this book to discuss the principles of modularization in detail, but we present a short summary of the general principles.

Put simply, the good product programme has two properties, namely a suitable variety seen from the market, and a suitable commonality seen from the company's point of view. The figure below shows an example of modularization for a washing machine and a drier.

The washing machine and tumbler drier are built with modules used in both products. The drum and basic chassis are common to both products. Depending on whether it is to be a washing machine or a drier, soap dishes, controls, coloured facings etc. are added. Seen from a production point of view, this means that the creation of variants takes place at a later stage in the process than it did before modularization. This means that the basic modules can be stored and assembled into a washing machine or a drier, depending on the customer's current order. This gives shorter delivery times and a more uniform production flow. All things considered, this achieves a high degree of commonality within the company, while from point of view of the customer, it has become possible to create more customer-specific variants, adapt the products to current customer needs, and thus achieves higher prices.

This example shows that there is not necessarily a conflict between a high degree of variety and a high degree of commonality.

The basic mechanism in modularization is encapsulation of complexity in the modules. This means that when the modules are used in product development and configuration, then properties, such as function, production, service etc. are known. Many investigations have shown that the complexity in the product programme creates complexity in the entire company with considerable overhead costs as a consequence. An American investigation [Guess, 2002] has surveyed the relationship between the quality of the data that the company works with, and the resources that are used to create, deal with and use product data (cf. figure 5.2).

If the data are 100% correct, then the resource consumption needed for performing the activities has, say, index 100. The investigation shows that if the correctness of the data falls to 97%, then the resource consumption has index 125. And if the correctness falls to 92% then the resource consumption has index 200. This means that half of the company's staff is busy with activities that do not create any value. Thus, there is every reason to evaluate the product programme and the correctness of the relevant data critically before a configuration project is started.

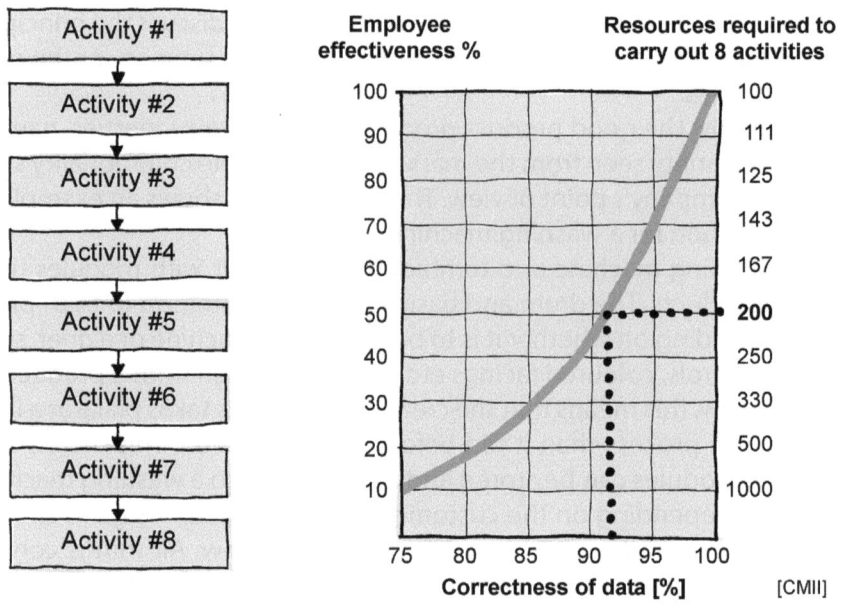

Figure 5.2. *Relationship between data correctness and resource consumption [Guess 2002].*

Using a configuration system can contribute to improving the quality of product data – partly because a configuration project leads to "tidying up" product data and rules, and partly because using a configuration system helps avoid many human errors so that the product data produced by the configuration system (such as offers, lists of parts and lists of operations) are more free of errors.

The company Scania has, as mentioned previously, over a considerable period of time created a very competitive product programme via a systematic use of modularization. Scania [Scania, 2004] gives the following reasons for working strategically with modularization:

- "A customer pays more for a highly specified vehicle than for a standard product"
- "With Scania's modular system the customer can specify the vehicle that he/she wants"
- "The more closely vehicles can be adapted to a transport task the better the customer's operating economy will be"
- "The modular system is important to Scania's development, production and product quality"

- "It simplifies parts management, contributes to higher degree of service"
- "The modular system provides a carefully balanced number of main components with great flexibility"
- "This allows considerably longer production runs than is possible in a conventional product system"

[Ericsson and Erixon 2000] have identified twelve drivers for modularization. These drivers can provide the rationale for placing interfaces between the modules in a product range in various places. The twelve drivers are:

- *Carry over:* This driver refers to modules that are able to be used unchanged in future generations of the products. An example could be a power supply where variants do not directly create any value for the customers and can therefore be used unchanged.
- *Technological evolution:* This is a question of encapsulating what is technologically constant and what is subject to change. For a DVD loader, it could be relevant that electronic and mechanical parts are strictly separated, since the mechanical part will often be more durable than the electronic part of the product.
- *Planned product changes:* With this driver, an attempt is made to create modules which bear important properties of the product. For example, a piece of measuring equipment, could be prepared with interfaces so that a measurement module which was not specifically known at design time could be installed directly into future equipment.
- *Technical specifications:* Solutions with technical choices that is significant for creating variation in such technical specifications as function, size etc. can be encapsulated in modules. An example could be gearboxes for cars, which are designed with the same physical interfaces but vary in functionality.
- *Styling:* With this driving principle, the product is made robust with respect to changes in such aspects as fashion and trends. An example could be an office chair designed with an internal skeleton that can be decorated with design shells that follow fashion and trends.
- *Common units:* The driver here is that the same modules are used in several products. An example is undercarriage platforms in the car industry, which are shared by several models.
- *Process and organisation:* The rationale for this type of modulariza-

tion is knowledge inside and outside the organisation. An example could be that everything related to sound reproduction is managed in a module for which a sub-contractor has developed state of the art competence.

- *Separate test:* With this driver, the products are comprised of modules that each has a well-defined functionality which can be tested separately. Then, the final test only needs to test the entire product. An example would be software routines which can be tested separately.
- *Black box deliveries from suppliers:* With this driving principle, subcontractors develop, produce and deliver modules, whose interfaces are well-specified and well-documented. Such modules are also known as assembly ready units.
- *Service and maintenance:* Here the products are designed so that servicing conditions for them are optimal. An example would be a control panel which is not repaired if defective, but instead just replaced by a new module.
- *Upgrading:* With this driving principle, the products are structured to allow for later upgrading. An example is PCs, where it is possible to upgrade the hard disks, memory etc.
- *Recycling:* Here the reason for the division into modules is to optimize the conditions for recycling. An example is copying machines, where certain modules are renovated after disposal of the entire product. After being renovated, the modules are inserted into new products.

In modularization projects, the twelve drivers will often appear in combinations where several of the drivers overlap each other. Experience shows that to obtain good effects from modularization, it is essential to consider which business benefits are the important ones - for example, reducing time to market, achieving greater efficiency in development efforts, opening possibilities for multi-sourcing, reducing costs etc. The twelve drivers listed above can be used as thought patterns during development, and as a checklist in connection with the search for solutions.

In this book, modularization is not discussed in more detail, but it is assumed that the product programme is ready to be used in a configuration system.

Theoretical background for analysis of product range

This section presents a short introduction to the theoretical background for analysis of a product range. Three important areas of theory form the foundation for this: system theory, object-oriented modelling and multi-structuring.

- System theory can be used as the basis for modelling the components of a product range or family. The strength of system theory is that there is a sharp division between what the product can (function) and what the product is (structure).

- Object-oriented modelling offers a lot of important mechanisms for modelling a product range. The most important principle is classification using generalization–specialization relations. When a product range is modelled using object-oriented modelling, it becomes possible to see what is similar and what is different over the entire range. A further strength of this form of modelling is that not just a single product but a whole product family or range can be described in an extremely compact manner. In the following discussion of product variant masters, classification is examined more closely. Object-oriented modelling is described in chapters 3 and 6.

- Multi-structuring assumes that it is relevant to model the product from several points of view. In the context of configuration, at least three viewpoints are relevant: the customer, development and production views, which correspond to important stakeholders in configuration.

A more detailed description of system theory and multi-structuring follows.

System modelling

The word system has at least two meanings – one is used in common speech and the other is more precise. In common speech, expressions such as "the system is unfair" or "it is the system's fault" are used. In this work, we use the term system in a more precise sense. According to system theory, [Haberfellner, 1994], a product or product family can be considered as a collection of elements and relations. The system is clearly separated from its environment and interacts with it. The system receives input from its environment and creates output for its environment. Figure 5.3 illustrates a system.

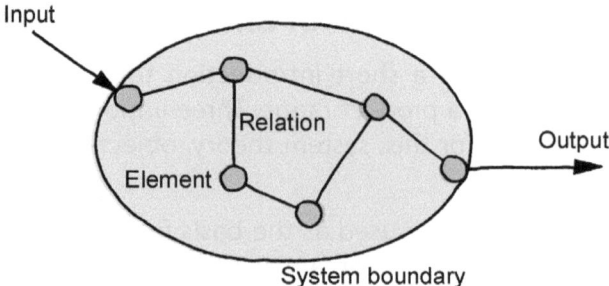

Figure 5.3. *A system.*

Example

If a car is considered as a system, the elements can be engine, chassis, brake system etc. The relations will be spatial and functional – so the engine and chassis are spatially related while functionally petrol flows from the tank to the engine. Input, for example, is petrol, and outputs are the transportation of people and goods. The car's environment is the atmosphere, the roadway etc.

In the context of creating a product model in a configuration system, it is important to consider carefully the interface between the product model and the environment. Some of the questions which are often considered are: Are accessories to be included? Are manuals parts of the configuration system? How is software dealt with? The environment will often play an important role – for example, legal requirements may have an important influence on which configurations are permitted.

When something is considered as a system, two characteristics are important: its function and its structure. Put simply, structure answers the question, "What is it?"; function answers the question, "What can it do?". In the example of the car above, the structural elements are the engine, chassis and brake system, while the functional features are top speed, noise level etc. The distinction between function and structure is important, because only structure can be determined directly. The function can only be derived when the structure has been determined. In the example with the car, the car's ability to drive with a certain speed (a function) is derived when a series of structural elements are known, such as engine capacity, friction between the wheels and the road etc.

In the context of building up configuration systems, it is a good idea to differentiate between structure and function. This basically means clarifying what can be directly determined by the user, and what are derived features, i.e. function. Often a configuration system will have both structural and functional features selected by the user. For a configuration system created for configuring cars, the structural features selected could be engine capacity, colour etc.

Functional features are top speed, acceleration, petrol consumption etc. Put simply, the more the configuration system is built up to be controlled by customer needs, the more functional features are specified, while the more the configuration system is oriented towards specialists, the more features are specified structurally.

An example where both situations are supported is the Dell configuration system. A computer can be specified in two ways: One way is to give the size of the hard disk, size of the memory, screen etc. The configuration system then creates a list of parts with the components to be used. In this case, structural features are specified. Another way is to state what the computer is to be used for, such as games or CAD graphics. With applications as the starting point, the configuration system creates a list of parts with the main components. In this case, the PC is configured according to the customer's functional requirements.

Configuration systems will often contain a mixture of functional and structural features.

Multi-structuring

Within the different functional areas of a company, many different ways of describing a product range are used. [Andreasen et al. 1996] have shown that for each phase in the lifecycle, one or more product structures exist that are relevant for product modelling. Figure 5.4 illustrates a product range described in relation to some different aspects of the life cycle.

Some examples of multiple structures: In connection with development of a product, it is normally most relevant to consider the product from a functional and structural point of view, where the components (structures) are functional elements in the product, such as a engine, a brake system etc.

Seen in relation to sales, it will be most relevant to know which functions, properties and degree of customer value can be created for the customers. Here, the product model consists of "sales features", which are relevant in connection with selling the product. From a production viewpoint, it is most relevant to consider the physical parts, i.e. the components created in the production and assembly processes. Correspondingly, shipping is of interest in a product model describing the parts that can be transported, for example so they fit in a 40-foot container.

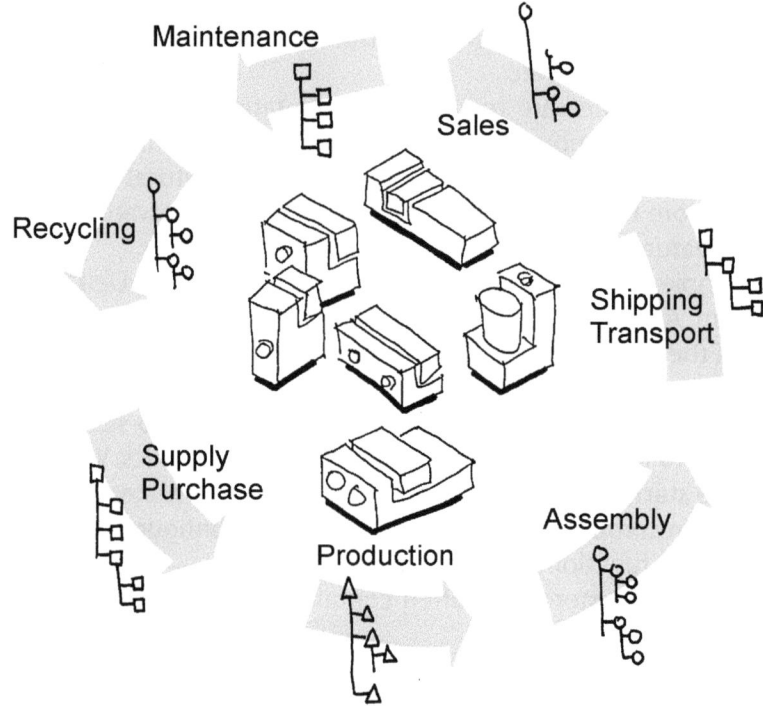

Figure 5.4. *A product range has various structures [Andreasen et al., 1996].*

One might now ask which structure the product model in a configuration system must have. In our work with implementing configuration in companies, we have found it practical to describe the product range from three points of view:

- *Customer view.* Here attention is focussed on the product range's functions and properties, and on the structures relevant for the customer.

- *Engineering view.* Here attention is focussed on the relationship between the product range's functions and structures by using solution principles.

- *Production (part) view.* Here attention is focussed on the product range's detailed structure and the life-cycle properties of production and assembly.

The three areas can be said to reflect the primary stakeholders in the configuration system, irrespective of whether it is a question of front office or back office configuration systems. A brief description of the three views is as follows:

Customer view contains the aspects of interest to the customer. Normally, this concerns functions and properties of the product, such as performance and interfaces to the environment. The information base is contained in existing sales systems and the knowledge of sales staff and product management. In simplified terms, the customer view must make it possible to explain what makes customers buy the products.

Engineering view contains the functional units and solution principles contained in the product or product range. The information base is the engineering systems and applications, such as CAD, PDM etc., together with the knowledge of the designers and product developers. The engineering view must make it possible to explain how the product works, and which functional variants exists.

Production (part) view contains the physical components to be dealt with in production and logistics. The information base comprises existing ERP systems, together with knowledge possessed by production, purchasing and logistics staff. The production view must make it possible to explain how the product is produced, based on output from product development.

With a slight exaggeration it can be claimed that the individual customer, development and production views taken separately may be uninteresting but that good business is created when the right relations exist between the three points of view. In order to have the relevant knowledge in the configuration system, it is quite essential that all three points of view are represented in a configuration project. Depending on the type of configuration system to be established, the weighting between the three will differ.

The three theoretical foundations – system theory, object-oriented modelling and multi-structuring – can perhaps seem a little abstract, but in the context of product analysis, they have turned out to be useful in connection with obtaining a general view of a product range and determining its scope. Now that we have presented this theoretical basis, the next section describes the primary tool for product analysis, the so-called product variant master.

Product variant master

In this section, the product variant master and the associated conceptual apparatus are presented in detail. The terminology is based on [Harlou, 2006]. Before going into detail, we present a little example of a product variant master: Figure 5.5 shows a product variant master for three families of cars.

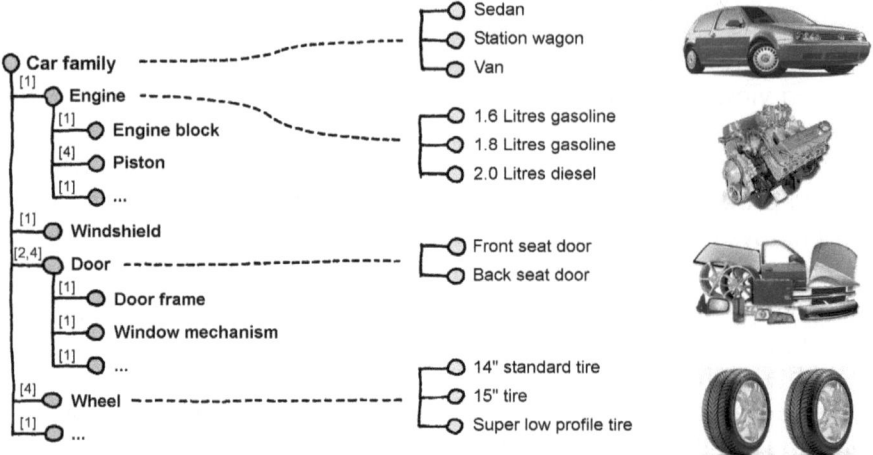

Figure 5.5. *Product variant master for three families of cars: sedan, station wagon and van [Harlou, 2006].*

The left hand side of the figure describes the products' generic structure. The car comprises engine, windscreen, doors, wheels etc. The right hand side shows the variants available. There are three families: sedan, station wagon and van. Similarly, the figure shows that the engine is available in three variants: 1.6 Litres gasoline, 1.8 Litres gasoline and 2.0 Litres diesel.

In general, the left hand side (known as the part-of structure) in a product variant master shows the components included in the products, while the right hand side (known as the kind-of structure) shows the variants available.

In the following sections, we describe the modelling principles for the product variant master in detail. The description may sometimes seem over-detailed, but if work with the product variant master is divided among several people, then it is important that there are fixed principles that everyone follows. The following "language" for modelling products and product families is general for all products and can describe a whole product range, its complete range of variation, and the total amount of knowledge that exists about combining its parts.

Class definition

A class is defined in accordance with the object-oriented paradigm as a group of objects with the same structure or function. In a product variant master, each class is marked with a filled circle. An example of a class is wheels, which can consist of subclasses such as large and small wheels. All of the wheels, i.e. the class wheels, have descriptive attributes such as size, materials, colour etc., which are common for all wheels.

Each class has a unique name in order to avoid misunderstandings in the interpretation of a product variant master. Figure 5.6 illustrates how classes are described with circles with a short horizontal line on the left. The most fundamental attribute of a class is its name. The name of the class is written to the right.

For practical reasons, the class name should be relatively short. When there is a need for more explanation, a description is used. The text in the description field is written in a grey font, cf. figure 5.6.

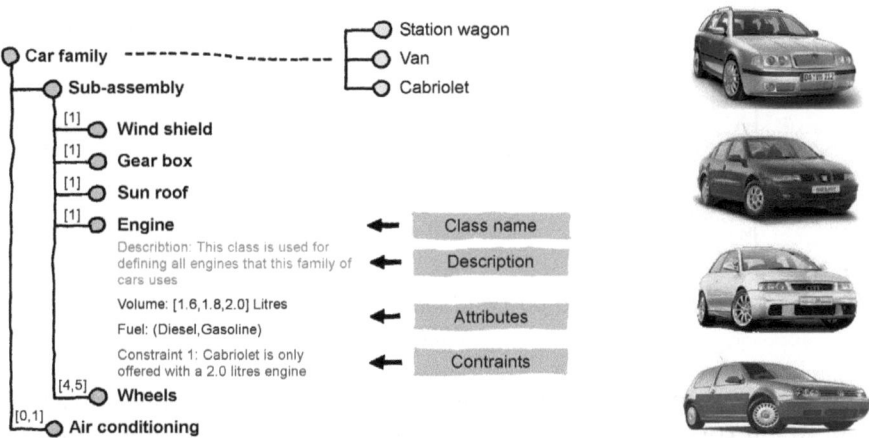

Figure 5.6. *Principles for describing classes in a product variant master [Harlou, 2006].*

Classes can contain attributes. These attributes describe the class and its attribute variations. An example could be the attribute colour, which can take on the values blue, green and yellow. The attributes are written under the description field, or if there is no description directly under the class name.

Constraints describe rules for how classes and attributes can be combined. Constraints described in text are placed under the list of attributes.

Class hierarchy – part-of and kind-of structures

Classes take part in two hierarchies, which are known as part-of and kind-of structures. As shown in figure 5.7, a class can contain one or more classes in a part-of structure,. If we read upwards, the class on the higher level is denoted a super-part. If we read downwards, the class is denoted a sub-part. In the example below, the class "Car family" is a super-part with respect to the class "Subassembly". The class "Wind Shield" is a sub-part with respect to the class "Subassembly".

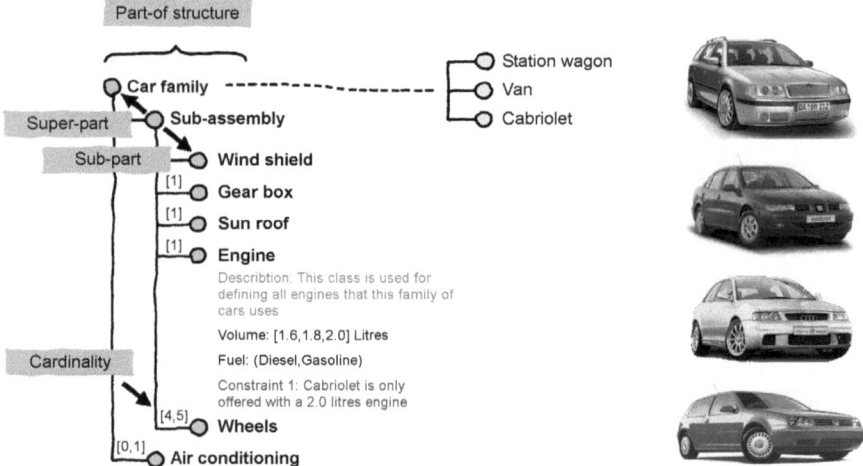

Figure 5.7. *Illustration of sub-part and super-part relationships in a class [Harlou, 2006].*

A class in part-of structures has cardinality. The cardinality describes how many sub-parts it has. For example, a car normally has 4 wheels; therefore, 4 is the cardinality for the class wheels. The cardinality is given by a number, for example [4], an interval [2-7] or a particular set of numbers [2,5,6]. In the example in figure 5.7 above, the cardinality for wheels is 4-5, which means that there can be 4 or 5 wheels in a given car.

Kind-of structures describe the range of variation for a given class. The relationships between classes in kind-of structures are denoted super-kind and sub-kind, as shown in figure 5.8.

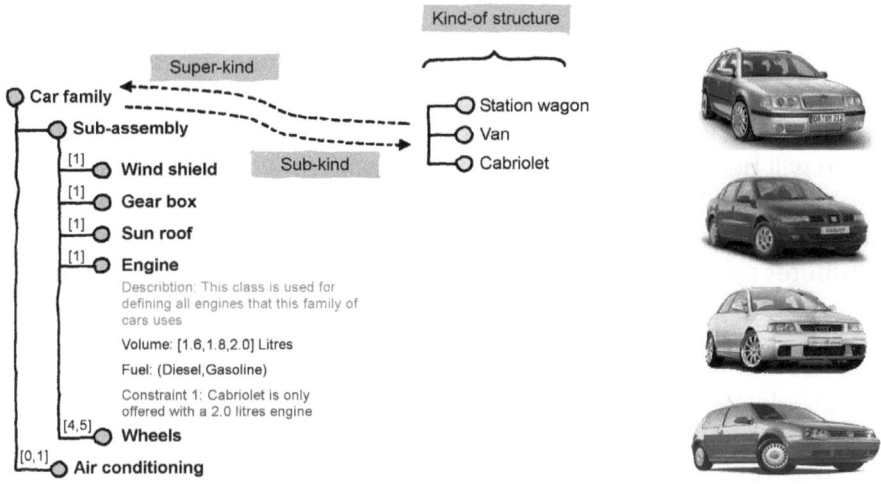

Figure 5.8. *Indication of super-kinds and sub- kinds for a class [Harlou, 2006].*

Attribute definition

Attributes are the descriptive parameters for classes, such as weight, price, code number etc. In a product variant master, one can distinguish between different types of attribute – for example an attribute can be identifier, real, integer or boolean. The significance of these is as follows:

- Identifier – An attribute described by a text string, for example colour [red, green, blue].
- Integer – An integer is a whole number and can be positive, zero or negative, for example: -10, 5, 0, 1, 2, 3. Integers can be specified as numbers [2], intervals [2-7] or a particular set of numbers [2,5,8].
- Real – A real number can be specified, for example, as 2.7, an interval [2.5 – 2.9] or a particular set of numbers, such as [2.7, 5.6, 9.4].
- Boolean – An attribute can be true or false, for example the existence of a hard disk [true, false].

Attributes are part of the class description, as shown in figure 5.8. An attribute is formally declared with an attribute name, followed by a value interval and a measurement unit: for example "weight [20-77] [kg]". Weight is the name of the attribute, [20-77] is the value interval, while [kg] is the measurement unit.

Constraints

A product variant master will normally contain many classes or components, which can be combined in different ways to form particular products. Often, it is not of interest to offer all combinations to the market, and there will be physical limitations on the ways in which components can be combined. To describe these relationships between components, constraints are used. A constraint is a rule which describes how classes and attributes can or cannot be combined. Four types of constraints exists:

- Verbal – Constraints are described by one or more sentences, such as "an open sports car is not sold with a sun roof".
- Logical – Logical constraints are explained with formal or semi-formal logic, e.g. "sports_car -> NOT sun_roof".
- Calculations – The constraints are given in terms of calculations, for example "car_weight=engine_weight + chassis_weight".
- Combination tables – Tables are used to describe relationships.

The table below describes relationships between parts and assemblies. The numbers in the table describes the number of parts used in an assembly.

A constraint is always declared in a class, even if it applies to several classes. Constraints are given a unique number or name as identification, cf. figure 5.9.

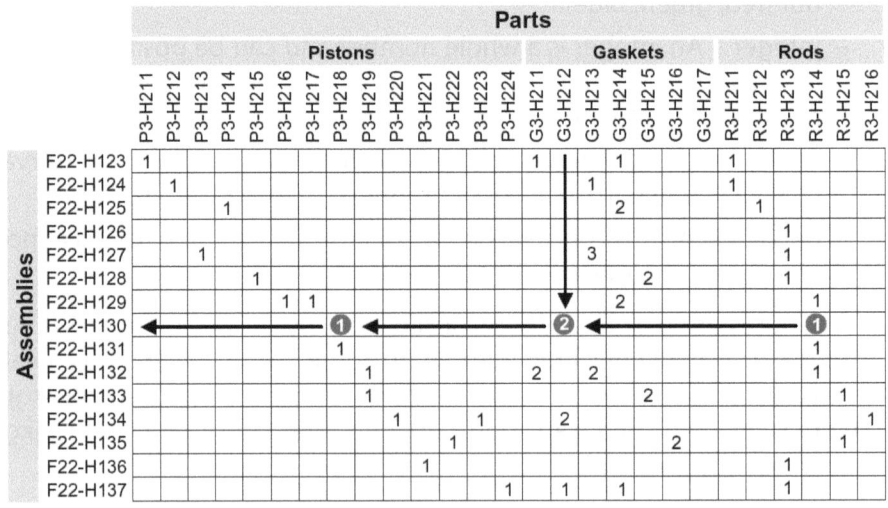

Figure 5.9. *Example of a combination table [Harlou, 2006].*

The presentation above describes the general formalism for a product variant master. Next, we look more closely at the individual views, i.e. the customer, engineering and production views.

Modelling formalism for customer view

The objective of the customer viewpoint is to describe those aspects that have the customer's interest. The range of variation modelled in the customer view is the commercial variation. The primary interest is the product's value-creating functionality, but there can also be other significant features. Structural aspects can also be relevant, such as the type of hub on a car. This means that a customer view contains both functional and structural aspects of the product range.

The modelling aspects relevant for the customer view are:

- Process modelling – that is to say, transformations of objects from one state to another. This means that what is modelled is what the product does or what the life- cycle systems do to the product. Examples of processes performed by the product include heating, cooling and combustion. Examples of processes performed by the life-cycle systems (the production system, logistic system, service system) include assembly, purchasing, servicing, disposal etc. (cf. the framework for modelling of product families described in chapter 2).
- Interface modelling – the interfaces between the product and the surrounding system can likewise be included in the customer view.
- Feature modelling – features are the structural and functional aspects that are relevant for a customer in connection with purchase, use and service of the product.

Process modelling

In process modelling, the way in which the product is used in interaction with the user and its environment is described. Process modelling is here based on the Theory of Technical Systems [Hubka, 1988]. A technical process system consists of four sub-systems, a technical system (the product), human (human system), the environment and the technical process. These four systems are illustrated in figure 5.10 and 5.11.

The line of thinking in process modelling is that the product, the environment and the user, together, create effects that make a technical process run. In the technical process, an operand is transformed from one state to another. An example could be the car considered previously. The technical process is transport of people and goods from one place to another. People and goods are here the operands. Making the process run requires the presence of a user (human system), the car (technical system) and a roadway (environment).

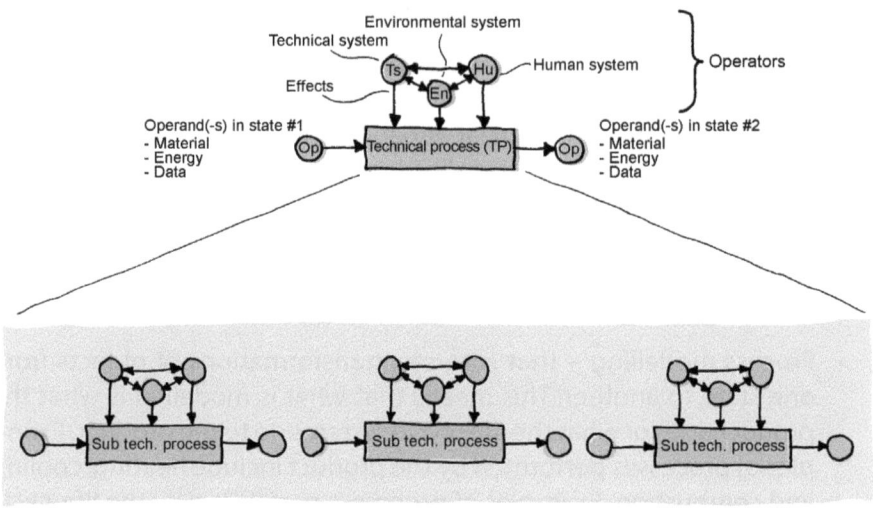

Figure 5.10. *A technical process system. A process can be broken down into sub-processes. [Hubka, 1988].*

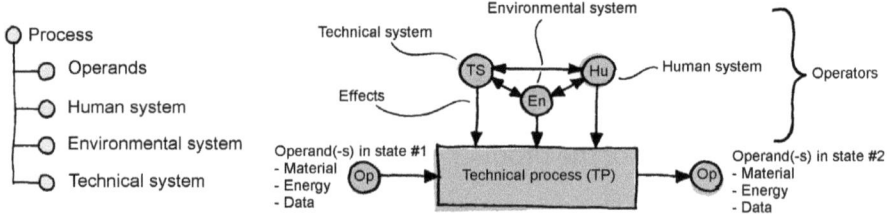

Figure 5.11. *Formalism for modelling processes. [Hubka, 1988].*

Process modelling can be performed with all four sub-systems, or just by describing the actual technical process.

Interface modelling

Another way of modelling the customer viewpoint is interface modelling. This means that the products' interfaces and their variation are modelled. This is especially relevant for products used in other products, such as valves, thermostats, controls etc. Examples of interfaces for pumps could be frequency, voltage etc. Such information is important in order to incorporate the pump into different applications.

Feature modelling

A third way of modelling the customer viewpoint is by describing features and their variation. Examples of a car's features are colour, navigation equipment, price etc. Features are modelled in both part-of and kind-of structures. Figure 5.12 shows an example of the customer view for a pump where the three modelling types for the customer view are combined.

The three modelling types can often be combined as shown below.

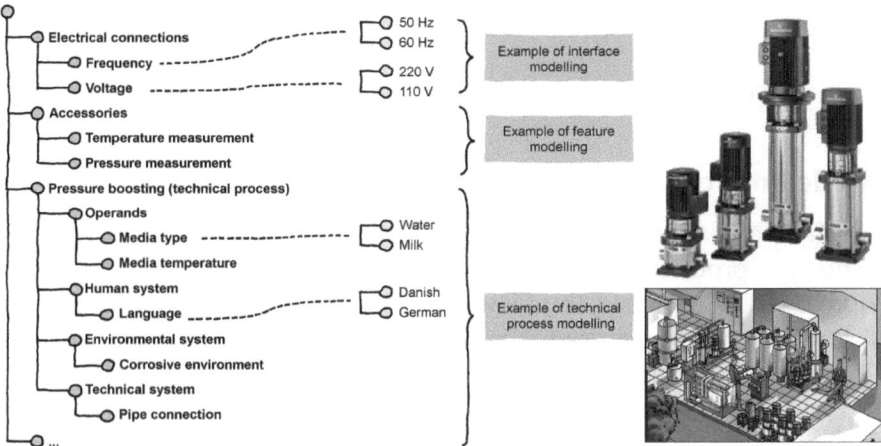

Figure 5.12. An example of a customer view for a family of pumps [Harlou, 2006].

Modelling formalism in the engineering view

From this viewpoint, the aim is to model the way in which the product works and the variations in this. This means that the aspects of the product range in focus are functions and relationships between functions and structures. Normally, the starting point for modelling in the engineering view is to identify the product's main functions and then identify the subfunctions. One of the main reasons for modelling from the engineering view is to acquire an overall understanding of a product range. This can be difficult in the production view, where for the more complex products there can be 10,000 or more components. The parts which are modelled in the engineering view are the functional units (organs) in the product, also known as solution principles.

Figure 5.13 shows an example of an engineering view.

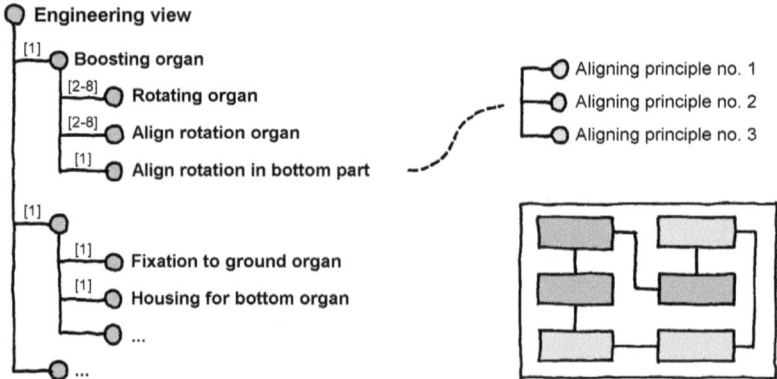

Figure 5.13. *Engineering view for a pump family [Harlou, 2006].*

Modelling formalism in the production (part) view

The aim of modelling from the production (part) viewpoint is to describe the product's physical components, which are either produced or purchased. In principle, this view is identical with a complete set of parts lists, which can be pictured as all the parts lists laid on top of one another. The modelling technique follows the part-of and kind-of structures.

Figure 5.14 shows an example of a production view for a pump family.

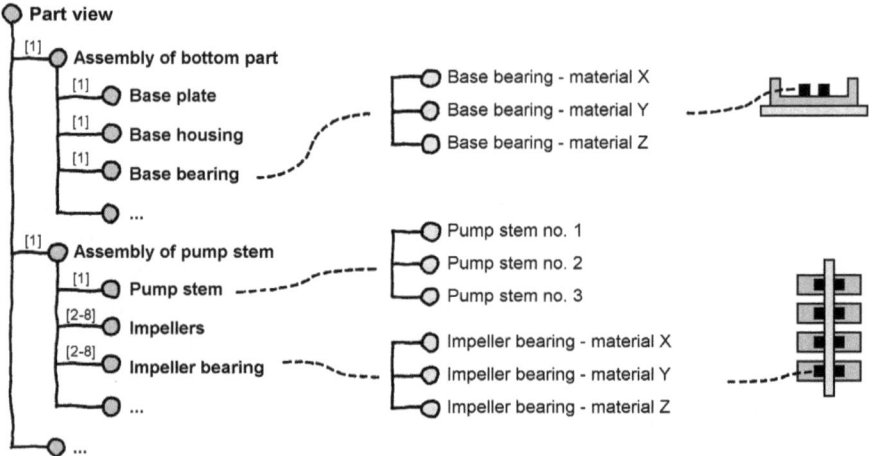

Figure 5.14. *Production (part) view for a pump family [Harlou, 2006].*

Relationships between the three views

A very important aspect of a product variant master is the relationships between the three views. These relationships are formally known as causal relations [Andreasen, 1996]. Such relationships exist between the customer view, engineering view and production (part) view.

The customer and the engineering view are related in such a way that customer features are realized by means of functional units in the engineering view. The engineering view and the production (part) view are related in such a way that functional units in the engineering view are realized by parts or interplay between parts. These relations are shown in figure 5.15, the left section as vertical lines.

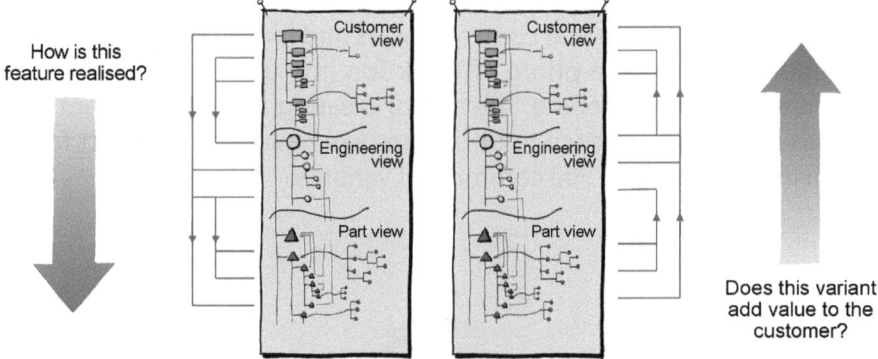

Figure 5.15. *Relationships between customer view, engineering view and production (part) view [Harlou, 2006].*

Correspondingly the production (part) and the engineering view are related in such a way that parts contribute to realization of functional units and thereby functionality. The engineering view and the customer view are related, thus functional units contribute to realization of customer features in the customer view. These relations are shown in figure 5.15, the right section as vertical lines.

In addition to the product variant master being able to secure knowledge about the product range, it also makes it possible to diagnose the product range. We discuss this further below.

Overview of value-creating variety. It is possible to see from the product variant master how large a proportion of the variants create value for the customer. The functional units, which in the engineering view cannot be related to the customer view, do not create value.

Complexity in the product range. The relations between the customer view, engineering view and production view offer us the possibility of evaluating the product programme's complexity. The number of relations between the different views indicates the degree of complexity in the product range.

This means that when changes have to be made in the product range, the magnitude of the task to be performed is roughly proportional to the number of sub-systems that have to be dealt with. Change and updating of sub-systems introduces risk and creates new tasks which have to be performed in production, purchasing, logistics etc.

Example of a product variant master

This section describes an example of a product variant master for the company Alfa Laval. The product is a so-called "Think Top". A Think Top is a mechatronic product that controls valves and is used in the food industry. Figure 5.16 shows a photo of Think Top (the uppermost unit), which measures the valve opening. The aim of creating a product variant master was in the first instance to obtain an overview of the product family, which includes around 500 commercial variants.

Figure 5.16. Left: Think Top mounted on a valve. Right: Discussion of the product programme based on a product variant master [Harlou, 2006].

Customer view

The approach chosen here for modelling the customer view is the interface modelling method described earlier in this chapter. The relevant factor here is the type of valve on which Think Top is to be mounted and the associated interfaces. Figure 5.17 shows part of the customer view for the product family.

The part-of structure contains: valve type, sensor interface, magnetic valve interface, air interface, electric connections etc. Variants of interfaces are modelled with kind-of structures. For reasons of confidentiality, the kind-of structures have been anonymized.

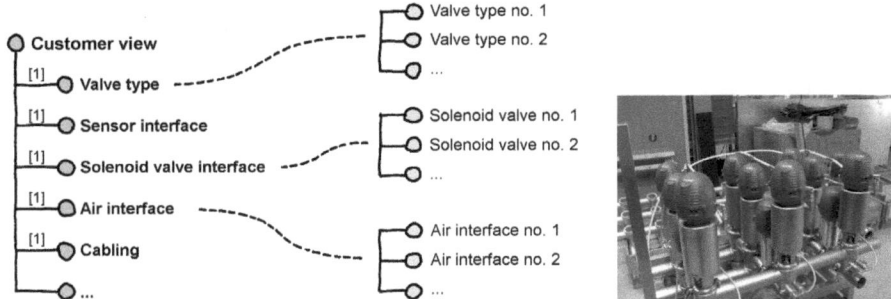

Figure 5.17. *The principle structure for the customer view [Harlou, 2006].*

Engineering view

The engineering view reflects the functional structure of the product range. In the part-of structure, the adapter module, basis module, sensor system etc. are modelled. Figure 5.18 shows part of the engineering view for Think Top.

The kind-of structure shows variants of the modules.

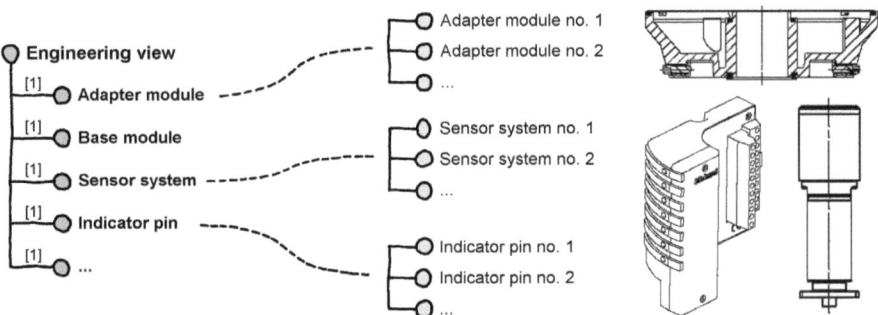

Figure 5.18. *The principle structure for the engineering view [Harlou, 2006].*

Production (part) view

The part-of structure for the production view is identical to the engineering view because of the functionally divided modular structure. The production view is, however, more detailed, since all the physical components are included. Figure 5.19 shows the principle structure for the production view.

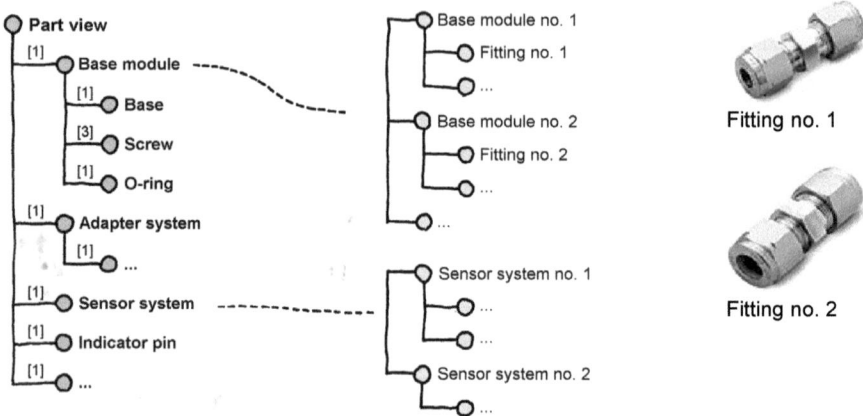

Figure 5.19. *The principle structure for a production view [Harlou, 2006].*

Relationship between the views

The three views are related to one another, as previously mentioned. Figure 5.20 shows how the customer, engineering and production (part) view are related. The causal relation are shown with vertical lines.

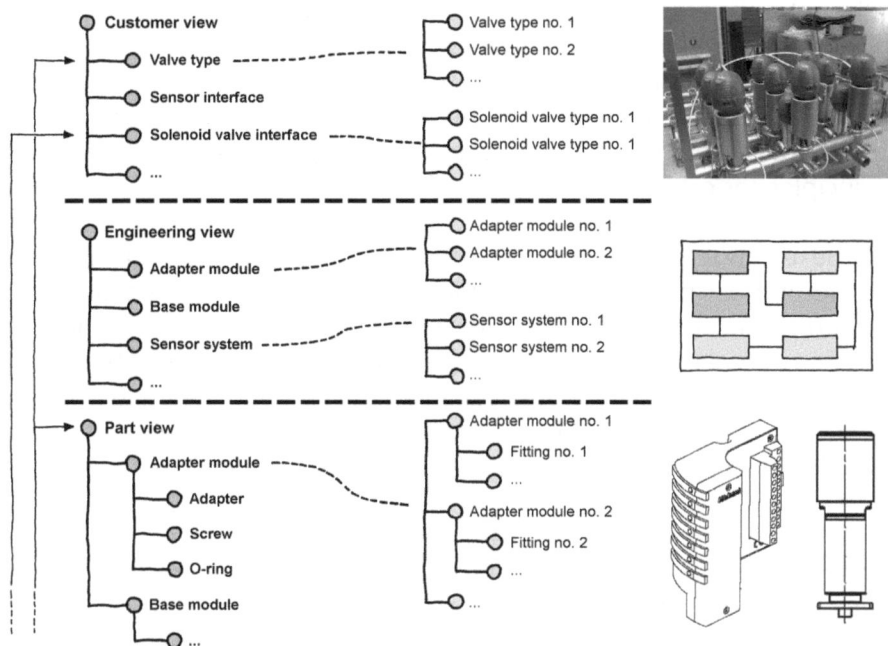

Figure 5.20. Section of a Think Top product variant master showing causal relations (vertical lines) [Harlou, 2006].

Product range analysis at Doors Inc.

Following is a description of what happened during an exercise in product analysis by the company Doors Inc. After the business goals for the configuration project and the task at hand had been clarified in phase 1, the consultant and an experienced design engineer started the product analysis.

The first step was interviews of experienced sales staff and sales managers. The aim here was to accumulate input to create the content for the customer view in the product variant master. For each of the market segments at Doors Inc., the obligatory and the positioning features were mapped out. The obligatory features are those providing access to the market, while the positioning features are those that make the products better than the competitors'.

The consultant structured the inputs from the interviews following the customer view in the product variant master. An important aspect is the heat conduction through the door. The experienced designer collected relevant standards for calculating the heat conduction for the door.

After about 14 days' work, the content of the product variant master was reviewed by a reference group consisting of staff from sales, development, production and logistics. This showed that several details regarding the elements for attaching the door were missing. In addition, the principles for calculating the price of overhead doors were discussed in detail.

The review led to a heated debate on the justification for some features in the product programme. The sales staff's evaluation was that all the features were absolutely necessary to ensure competitiveness, while production and planning were able to produce information indicating that a considerable part of the product programme had not been sold in recent years. The conclusion of the review meeting was that the sales managers should consider whether all features should be included in the configuration system.

Once a first version of the customer view was finished, the consultant and the experienced designer started to create the engineering view in the product variant master. Important inputs for this were 3-D drawings, lists of design parts for the doors, and the design handbook.

The experienced designer, who had a background in mechanical engineering, could immediately create a skeleton for the engineering view. In the areas of software and electronics, he received assistance from colleagues. Together with the consultant, discussions were held about different ways of structuring the engineering view, and after several attempts, everybody agreed that a structure divided according to the door's main functions provided the best overall view.

After a first version of the part-of and kind-of structures had been created, the project group began identifying the most important rules for configuration of the doors. A number of the rules could immediately be taken from the design handbook, while others only became apparent after interviews with the department's most experienced staff. After about two weeks, the engineering view's content in the product variant master was ready for review by the reference group.

I connection with the review, it became clear that each order processor and designer had his own way of placing control components in the control box. Seen from a customer's point of view, there could be no market-related or technical reasons for placing them differently in different doors. One of the results from the review meeting was that a consensus was reached on a single way of placing control components in the control box, both within the configurator and elsewhere within the company.

The next step in the establishment of the product variant master was to determine the production (part) view. Important inputs here were the lists of parts in the company's ERP system. After about two weeks' work, a first version of the production view was ready, and its content was reviewed by the reference group.

One of the points discussed at the review meeting was the level of detail in the production (part) view description. It soon became clear that it was not a good idea to include all details, such as screws, washers etc. At the review meeting, consensus was reached that it was the most important price-related components that should be included.

The next step in building up the product variant master was to collect data for all components in the product variant master. For each component, data for dealing with heat conduction calculations, strength calculations etc. had to be collected from various data sheets and databases. The consultant and the experienced designer worked full time on this, and the content was reviewed every Friday with the reference group. After working three more weeks, a finished product variant master was available, the content of which had been checked with the reference group.

Figure 5.21 gives a rough view of the content of Doors Inc.'s product variant master.

In addition to the fact that the most important knowledge concerning the product range was now described on paper, this work contributed to creating consensus concerning the product programme. The reference group was surprised to discover how many variants of doors and sub-systems existed within the company. Similarly, it became more obvious to the sales staff what consequences the introduction of new variants had for development and production.

It was then decided to start work on defining Doors' preferred solutions, which to the greatest extent possible should be sold to the customers. It is Doors' impression that the product variant master has created the basis for a far more professional dialogue between sales, development and production. One designer expressed it in this way: "This is the first time we have had the possibility of a meaningful dialogue with sales".

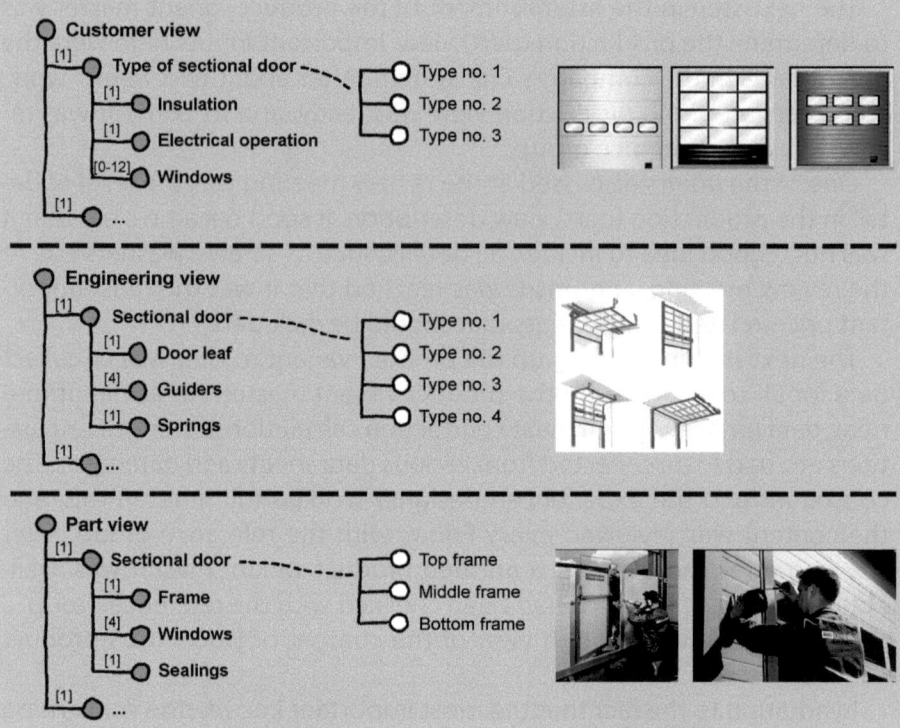

Figure 5.21. *Product variant master for Doors.*

Working method in analysing the product range

This section considers some useful principles in connection with creating the product variant master.

Think big – start small: It is important, even if to start with it is only a small part of the product range which is to be configured, that a structure for the whole is thought through. The product variant master can then be roughly sketched in its entirety, while the parts of primary relevance in the first stages of product configuration are structured in more detail.

Involve the stakeholders: A very efficient way of working is to have a small core group to create the product variant master and review it at intervals, together with relevant stakeholders from sales/marketing, development and production. The visual form of a product variant master makes it easier to involve experts. This becomes much more difficult once the knowledge has been described in databases and configuration systems. Use of a product variant master also makes it very obvious that progress is being made in a project. This can be measured in a very concrete manner, as the product variant master gradually grows larger.

Use existing data sources: Part of the content of a product variant master is to be found in existing IT systems, such as Sales systems, Product Data Management systems, Enterprise Resource Planning systems and Design handbooks. It is useful to use these systems as a starting point, but they are normally not sufficient. A large part of the constraints normally exist only in the heads of experienced sales staff, designers and the like. This means that a process to formalize the silent knowledge possessed by relevant resource persons is usually necessary. If the majority of knowledge has to be identified through interviews, the effort needed to create a complete product variant master can be considerable.

Consider the system boundaries: It is important to identify the right interface between the product and the environment. Some of the aspects to be considered are the extent to which installation manuals, government regulations, building parts etc. should be included in the product variant master.

Consider maintenance of the product variant master: In connection with the task of developing the product variant master, the maintainability of the configuration system (or the lack of it) are determined to a considerable extent. It is important to consider which parts of the product variant master are expected to be stable over longer periods of time, and which parts are dynamic and therefore have to be maintained. In a mechatronic product, it could for example be the mechanical parts that are stable, whereas the electronic part has to be updated more frequently.

Use figures and diagrams: In order to be able to communicate the content of the product variant master to non-experts, photos, diagrams and drawings contribute to increasing the communicability of the product variant master's contents. Figures can also help maintain awareness of important reasons for creating the product variant master, such as the definition of concepts.

Consider the usefulness of the product variant master: In addition to the fact that the product variant master maintains knowledge about the product range, many other effects have resulted from its use. In many companies, the quality of the dialogue between sales/marketing, development and production is improved. Sales/marketing obtains a clearer understanding of the consequences for development and production of introducing variants. Development and production obtain clearer understanding that not all variants in the product programme necessarily create value for the customers. In the majority of companies, most functional areas have been surprised at the size and complexity of the product range.

Using software for drawing the product variant master: In connection with projects, we have experimented with various types of software to support drafting the product variant master. We ended up using MS-Excel, MS-Visio or AutoCAD. We defined a template to support drafting. It contained the most important components, such as part-of and kind-of relations. Particularly in large projects, where the modelling task can be divided among several people, the template helps make the product variant master more uniform.

Bibliography

[Fabricius, 1994]: Fabricius, F: Design for Manufacture DFM Guide for improving the manufacturability of industrial products, EUREKA, 1994.

[Hubka, 1988]: Hubka, V. & Eder, W.E.: Theory of Technical Systems, Springer-Verlag. Berlin 1988

[Meyer & Lehnerd, 1997]: Meyer, M, H, Lehnard A., The Power of Product Platforms: Creating and Sustaining Robust Corporations, Simon & Schuster, 1997

[Ericsson & Erixon, 2000]: Ericsson, A. & Erixon, G, Controlling Design Variants: Modular Product Platforms, American Society of Mechanical Engineers, 2000

[Andreasen et al., 1996]: Andreasen, M. M., Hansen, C. T., and Mortensen, N. H. "The Structuring of Product and Product Program", Proceeding of the WDK Workshop on Product Structuring, The Netherlands, Delft University of Technology, 1996

[Klir & Valch, 1965]: Klir, J. and Valach, M. "Cybernetic Modelling", Iliffe Books, London, 1965

[Scania, 2004]: Scania. "Årsedovisning 2004", [In Swedish], Sweden, 2004

[Harlou, 2006]: Harlou, U. "Developing product families base on architectures – Contribution to a theory of product families" Dissertation, MEK, Technical University of Denmark, 2006

[Haberfellener, 1994]: Haberfellner, R. et al. "Systems Engineering - Methodik und Praxis", [In German], Verlag Industrielle Organisation Zürich, 1994.

[Guess, 2002]: CMII for Business Process Infrastructure, Holly Publishing, Scottsdale, US 2002

6

Object-Oriented Modelling

We start this chapter with a short introduction to the object-oriented way of thinking. Then, we describe the basic concepts and the associated terminology in the object-oriented paradigm. The chapter is based on books about object-oriented modelling by Agida, Bellin, Booch, Coad and Yourdon, Graham and Rumbaugh. The notation used is taken from Unified Modelling Language (UML). The aim of the chapter is to introduce the basic principles of object-oriented modelling and to describe how these principles can be used when you develop a configuration system.

Basic concepts of object-oriented modelling

Object-oriented modelling is a method of analysis for modelling knowledge and information in a given system, where the system is broken down into sub-elements. The principle in object-oriented programming is illustrated in figure 6.1.

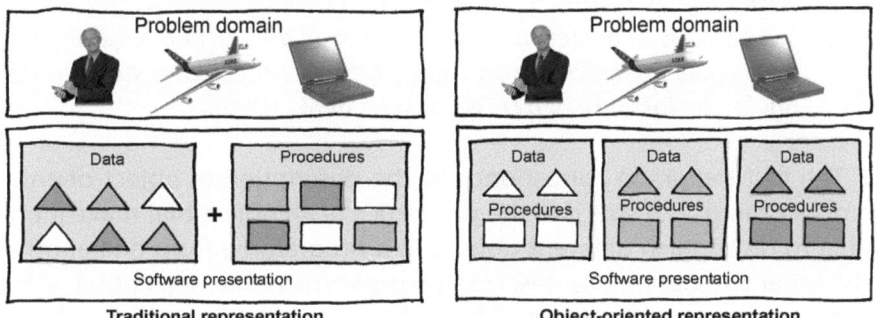

Figure 6.1. Structure in traditional and object-oriented software [Adiga, 1993].

The traditional development of software known as "structured programming" is based on the structured/functional paradigm, in which the problem domain is divided into subsystems. Each subsystem is associated with some functions or procedures, which operate on a common set of data. This structure is not very convenient if the system is to be extended, as it is often difficult to see whether any of the changes will have consequences for existing procedures or data. By using the object-oriented approach, we make this problem easier to deal with, as data and procedures are preserved as a unit collected around the objects that have been identified within the problem domain. This produces a module-based structure that is easier to extend or change.

Object-oriented modelling is characterized by objects being able to inherit information and procedures from other objects, and by the fact that the individual objects' information and procedures are encapsulated. From this starting point, we now introduce the basic concepts in object-oriented modelling.

Object

The basic elements in object-oriented modelling are objects. An object is both an abstraction of an element in a domain and an element in the software describe by an identity, a state and a behaviour. Below we quote three different definitions of objects, which together cover both objects defined in the domain and objects as elements in a programme:

a) *"Object: An entity with identity, state and behaviour."* [Mathiassen, 1998].

b) *"An abstraction of something in a problem domain, reflecting the capabilities of a system to keep information about it, interact with it, or both; an encapsulation of attribute values and their exclusive services. (Synonym: an instance)."* [Coad and Yourdon, 1990].

c) *"An object is a concrete manifestation of an abstraction; an entity with a well defined boundary and identity that encapsulates state and behavior; an instance of a class."* [Booch et al., 1999]

The first definition corresponds to the description of object-oriented programming, as it specifies the properties of an object that make it possible to distinguish different objects in a programme from one another. The other two definitions describe the properties of an object that an object-oriented analysis of a specific problem domain aims to identify and model.

Objects can describe phenomena in the world around us. When compared to language, the identities of objects are described by nouns (things), their states by adjectives (appearance, quality) and their behaviour by verbs (actions).

In this way, the object's identity is given by its name; the object's state is given by its attributes, and its behaviour by its methods. A method reflects a process which an object can perform. A method can for example be to use constraints, call of values, control of user interfaces etc.

Class

Objects can be grouped in classes. A definition of a class is:

"Class: A description of a set of objects with the same structure, behaviour patterns and attributes" [Mathiassen, 1998].

A class is a type of model or general description of objects with identical characteristics, which can be described in a common way. Every object's identity, behaviour and attributes are described in general terms in the class. But each of them has its own identity and its own concrete state and behaviour. The objects "Peugot 407", "Toyota Corolla" and "Volvo S40" can be members of the class, Cars. This class will then provide a template for a series of different cars, described by their identities, states and behaviour.

Instance

As can be seen from the definition below and the third definition of an object [Booch et al., 1999], the concept of "object" and of "instance" describe the same.

"An instance is a specific object or an individual example from a class (an instance is itself an object and therefore has an identity, state and behaviour)" [Arngrimsson, 1992].

For example, the class "Cars" has the attributes: motor capacity [1.8 or 2.0] liters and the colours [red, blue, green]. An instance of the class could then be an object with the name Volvo S40 with motor capacity 1.8 liters and colour blue.

Abstraction

By abstraction is meant:

"Abstraction is the principle which aims to ignore the aspects of a domain which are irrelevant in relation to the objective, so that it is possible to concentrate on those that are relevant" [Mathiassen, 1998].

In a concrete example of system modelling, we ignore (abstract from) the less relevant aspects, which in most cases is necessary in order to be able to model. Thus, abstraction is connected with the viewpoints and the objective used in modelling a domain.

Abstraction is something we all use in daily life: When a driver declutches, he abstracts from the internal construction of the clutch. The designer of the clutch has encapsulated the clutch's internals, so a motorist can use the clutch without knowing in detail how the clutch is put together. When a user of a word processor system uses the Edit command, he abstracts from the way in which the Edit programme is constructed. The programmer has encapsulated the internals of the Edit programme. As indicated by these examples, the concept of abstraction is closely associated with the concept of encapsulation.

Encapsulation

In the object-oriented paradigm, encapsulation covers the principle that an object's external functionality is independent of the object's internal data structure and functionality. This encapsulation means that external access to an object's attributes can only take place by communication using pre-defined messages that the receiving object can interpret and use to start suitable actions.

By the use of encapsulation, it becomes possible for example to change an object's internal construction, without this influencing the overall system's functionality. The principle of encapsulation also offers better possibilities for re-use of previously written programme codes, since a system developer, on the basis of a description of the individual classes' external functionality and input parameters, can relatively easily identify usable classes and thus increase the extent to which they can be re-used. In this way, a system developer can free some resources, which would otherwise be used to understand the internal functionality of objects from a specific class, for other purposes.

The concepts of abstraction and encapsulation are closely related, as abstraction concerning objects is stong only if the objects are also encapsulated. This reduces complexity, since the system developer does not need to worry about the internal structure of the object. The concepts of abstraction and encapsulation can be summarized as follows: abstraction focuses on the object seen from the outside, i.e. the characteristics which make it possible to identify the object as a separate part with its own identity, whereas encapsulation focuses on the object seen from the inside, i.e. the properties which give the object 'substance' (the object's structure and function).

Inheritance

"Inheritance is the mechanism which delegates properties from one object to another, or from a class to an object or from a class to a class.

- *A class inherits all or part of its description/definition from another more general class or classes.*
- *An instance (object) inherits all its properties (attributes and methods) from the class to which it belongs".* [Arngrimsson, 1992]

Classes can be structured in a generalization-specialization hierarchy consisting of superclasses and subclasses. In such a hierarchy, a subclass, in addition to the properties it defines itself, will also inherit defined properties from its superclass. If the subclass inherits defined properties from more than one superclass, we speak of multiple inheritance.

The subclass re-uses all the superclass' properties. This is denoted inheritance. For example the subclass "Opel" could inherit all properties from the superclass car (motor, colour, bodywork, chassis number etc.). From a programming point of view, this gives a more secure code with a high degree of re-use, while ensuring that later changes in properties of the higher-level objects are inherited by the subclasses. In the section "generalization structure" below, a further description of the principles of inheritance is presented.

Communication through messages

The dynamic properties in an object-oriented system are represented by exchange of messages between the objects in the system. Communication through messages is closely associated with the concept of encapsulation, as encapsulation implies that the objects' data attributes can only be changed or read by sending a pre-defined message to the objects. How the object changes or reads these attributes is the responsibility of the object. The message is processed by the object's methods, which can both change the object's state and cause the object to return a value, a text string etc.

A method can fetch or change an attribute, perform calculations, or send messages to other objects. Thus, the object's methods form the interface in relation to its environment. The message itself consists of an address (of the object or objects it is to be sent to) and an instruction, consisting of a method name and zero or more parameters.

Object-oriented analysis (OOA)

Object-oriented analysis is a method of modelling a given problem domain and an application area (the system's usage properties). The aim of the analysis is to achieve sufficient understanding of the problem domain and the application area to be able to model/describe relevant aspects of the desired future system.

The problem domain is that part of the real world outside the system which is to be administered, monitored or controlled. The application area is the organisation (department, people) that is to use the system in order to administer, monitor or control a problem domain. The object-oriented analysis phase (both for the problem domain and the application area) is used to describe the desired future system. We now examine separately the principles for analysis of the problem domain and the system's application area.

The problem domain

When you build up an object-oriented analysis model (OOA model) for the problem domain, the activities shown in figure 6.2 are performed. The figure describes the OOA model as made up of five layers. The individual activities (layers) can be thought of as different views, which together make up the OOA model. The activities are generally performed in the order in which they are listed, but can in fact be carried out in any order. In practice, the modelling task involves a number of iterations between the layers of the model.

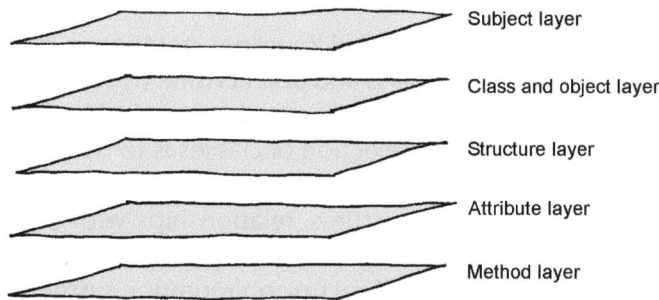

Figure 6.2. *OOA-modelling's five layers [Coad and Yourdon, 1990].*

- *The subject layer* contains a division of the complete domain to be modelled into various subjects. In relation to the use of product and product-related models, a subject can for example be a product model or a production model. The use of subject layers is analogous to the use of packages in UML notation.
- *The class and object layer* contains a list of the classes and objects that have been identified in the individual subject areas.
- *The structure layer* contains the objects' mutual relationships, i.e. a description of generalization-specialization and whole-part structures.
- *The attribute layer* contains a description of the information in the individual objects, i.e. what the objects know about themselves.
- *The method layer* contains a description of the individual objects' methods, i.e. what the objects can do.

The static structure is primarily reflected by the subject layer, the class and object layer, the structure layer and the attribute layer, while the more dynamic aspects in the model are mainly included in the method layer.

The various layers in the five layer model are described in the next section. The individual activities are presented in the following order:

- Find classes and objects
- Identify structures
- Identify subjects (package)
- Define attributes
- Define methods

Find classes and objects

By class and objects in Coad and Yourdons notation is meant "a class and the objects in the class". Class and objects directly reflect corresponding physical or conceptual features in a problem domain.

The primary technique for selection of classes is to write down a comprehensive list of candidates. When working out this list of candidates, you should be looking for structures, relationships with other systems, things or events which should be stored, description of roles, places, and organizational units. When building up configuration systems, the candidates for classes and objects are to be found in the product variant master, which is described in chapters 3 and 5.

The classes' names form the core of the language used to describe the problem area, so the terminology should be chosen with care. The aim should be to fulfil the following ideals: use simple and readable names, use notations from the problem area, use single nouns (if necessary, an adjective may be added), and use terms in the singular (for example car instead of cars). An indication of an unnecessary class can be a lack of attributes and methods [Rumbaugh, 1991].

Identify structures

Structures are used to describe the complexity of a problem domain and can either be generalization structures, package structures (which are used to describe subject layers), aggregation structures or association structures.

The structural relations can be divided into two levels: structures between classes and structures between objects. Structures between classes, which is the most abstract level, determine the conceptual relationships between two or more classes. Structures between objects determine the specific relationship between the individual objects. Thus, class structures describe static, conceptual relationships between classes, while object structures describe dynamic, concrete relationships between objects. Generalization structures and packages make up structures between classes, while aggregation structures and association structures make up structures between objects.

A recognized notation is from UML [Bennett et al., 2002], which is illustrated in figure 6.3. The figure shows the notation for a class and the four structures illustrated.

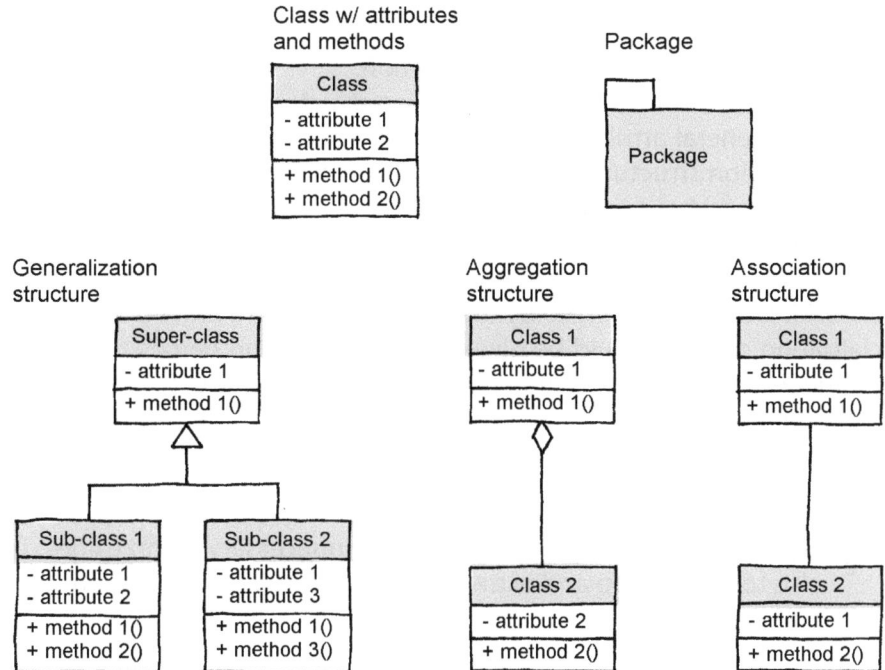

Figure 6.3. *UML notation for a class and four structures.*

Unified Modeling Language (UML) has been chosen as notation, because it basically builds on a separation between methodology and notation. There are no built-in rules for methodology in UML. Secondly, as a consequence of UML's large following, there is a large market for development tools that are compatible with the notation.

The aggregation structure and the association structure can be associated with a multiplicity/cardinality. The cardinality can be e.g. 0 to 1, 0 to * (* means an unlimited number), 1 to *, or exact values, such as 3.

The sign - / + in front of the attributes and the methods indicates whether they are private (-) or public (+). A public attribute or method can be read by all classes, whereas a private attribute or method can only be read by the class to which it belongs. Attributes are normally declared as private, in order to achieve encapsulation of the object, whereas methods are normally public, as communication between the classes/objects takes place via these.

Generalization structure

In everyday life, we often exploit possibilities for generalization. When we talk of "vehicles", this is a generalization of many different types of vehicle. Cars, motor cycles, lorries and buses are all specializations of a vehicle. In a problem domain, this type of structure is called a generalization structure or a generalization-specialization.

There is inheritance in a generalization structure, so the subclasses inherit the general attributes and methods from the superclass. Thus, the generalization structure is a structure, in which a general class describes properties and behaviour that are common for a number of special classes. These can, however, be extended in the specialization class, where it is also possible to add further attributes and methods. Linguistically, specialization can be expressed by the phrase "is-a" (e.g. a BMW is a car). Specialization classes should be given easily understood names that reflect the generalization class. When using generalization structures, it should be noted that:

- Everything can in principle be specialized, and it is important to watch out not to exaggerate the specialization effort.
- Try to make the model simple. Avoid unnecessary complexities. Try also to avoid deep and extremely complicated generalization structures.

An example of a generalization-specialization structure is shown in figure 6.4. The superclass is called "Car" and the subclasses are called "Volvo" and "Ford."

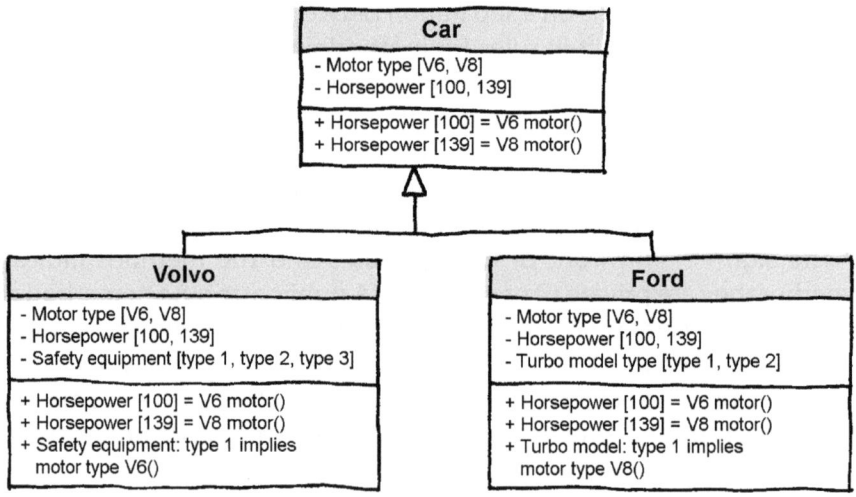

Figure 6.4. *Generalization-specialization structure*

The attributes "Motortype" and "Horsepower" and the methods from the superclass are inherited by the subclasses. New attributes and methods, which are not in the superclass, can then be added to the subclasses. In the figure, the subclass "Volvo" has had the attribute "Safety equipment" added, and the subclass "Ford" has had the attribute "Turbo model" added. Normally, inherited attributes and methods are only mentioned in the uppermost class.

Package structure

Package structure is the notation used to describe subject layers. Package structures are used to group classes, so that it becomes easier to get an overall view of the model. A package structure is a collection of classes which are mutually connected. The classes in a package are as a rule connected by generalization structures and aggregation structures. An example of packages is shown in figure 6.5.

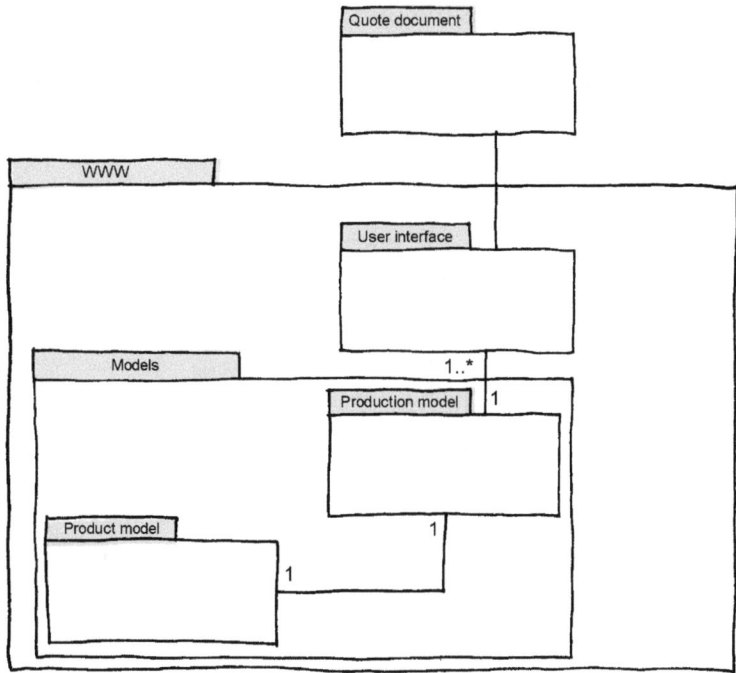

Figure 6.5. Packages.

Aggregation structure

This is a structure, where a superior object (the whole) consists of a number of subordinate objects (the parts). Linguistically, this sort of relationship is expressed by the phrases "has-a" and "part-of".

Aggregation structures (also called Whole-Part structures) can have the form whole-part, container-content or association-member, and are used to outline structural relationships between objects. An example of a whole-part relationship could be a car, which consists of a body and four wheels, while a container-content relationship could be a car, which contains a driver who is not a part of the car, and an association-member relationship could be a society with various members. The whole-part relationship is often easy to grasp, because this way of seeing structures is well-known from the bill of materials concept. Association-member is more difficult, since it requires more abstract thinking.

Figure 6.6 shows an example of an aggregation structure. In the example, the car class has three "subparts": body, motor and wheel, while for example the motor class has the car class as a "superpart". The car "has-a" motor and the motor is "part-of" the car. The cardinality is read as follows: each car has 4 or more wheels, while a wheel is part of only one car.

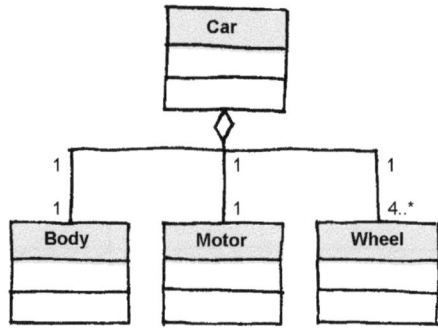

Figure 6.6. *Aggregation structure*

As previously stated, there is no inheritance in an aggregation structure.

Association structure

In this structure, a number of objects are associated with one another. Linguistically, we express an association structure by the phrases "knows" or "is-associated-with". An instance of en association structure is called a "link". A "link" is a particular connection between two objects (or classes) through which a message is sent.

An example of an association structure can be seen in figure 6.7. The car can have zero or more owners, whereas an owner can own one or more cars.

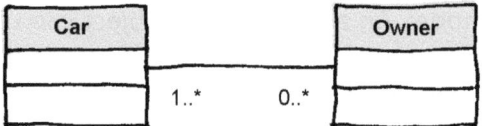

Figure 6.7. *Association structure.*

It can be difficult to distinguish between aggregation structures and association structures, as they have many common features. Generally speaking, an aggregation structure is a special, stronger form of association and can always be expressed as an association structure. If there is doubt about which to use, you can use only association structures, or you can use the aggregation structure to capture relations that are not at the same level and are more binding, and then use the association structure for those that are more on the same level and more transient.

It is important to evaluate systematically the structures that are being chosen. The structural relationships must follow these criteria: The structure types must be used correctly; the structures must promote understanding and an overview; they must be true to concept; they must be simple; and they must reflect the dynamics of the system.

Identify subjects (packages)

Subjects are used to group classes in a model to increase clarity. This is often desirable with large and complex models, where it can otherwise be difficult to get a clear overall view. Thus, subjects can be used to lead the reader through different subject areas in a diagram. Subjects can be found, for example, by promoting the uppermost class in each structure to be a subject, and also by promoting all classes that are not included in a structure to be a subject. In addition, it is possible to break down the problem domain into conceptual sub-regions and turn these into subjects. The subjects found should then be further developed, so that there are as few dependencies as possible in the form of generalization structures, aggregation structures, and association structures between the subjects. It is also possible to have subjects within subjects in large models. As previously mentioned, subjects are represented graphically by packages. Each package is described by using a tab with the package name (cf. figures 6.3 and 6.5). In large models, you can consider joining packages together to improve clarity.

Define attributes

An attribute is defined as data that has a value for all objects in a class, i.e. an attribute describes an object more closely within the chosen context. Thus, attributes are used as representation of states, which must be readable in the model. An attribute in an object can only be used by a second object if a message is sent from this object to a method in the first object. An example of an attribute is the attribute Safety Equipment in the class Volvo in figure 6.4. This attribute can take on the values type 1, type 2 or type 3.

To find the attributes, the following pieces of advice can be used:

a) Investigate how the object can be described generally and specifically in relation to the context for the desired system. Attributes describe the information which an object needs. At the same time, they are in an atomic form, which means that an attribute either has a single value or a close grouping of values.

b) The attributes are placed in the classes to which they are most closely related. In generalization structures, they are placed as high as possible.

c) Attributes should always have a meaningful value. Classes have as a rule more than one attribute.

d) Give the attributes easily understood names. Specify their value interval/range of results and units (e.g. kg, mm, kr).

Define methods

Methods are used to give a more detailed description of the behaviour which objects have the responsibility to carry out -- for example, changes in an object's state or communication between objects in the form of messages.

When methods are being defined, the following elements can be brought into play:

a) The methods can be divided into two types. One of these includes the simple methods, such as construct, connect, give access to, and liberate. The other type is the more complex methods such as calculate and monitor. The methods are identified by analysing the required behaviour for the object.

b) In order for an object to be able to perform its tasks, it can be necessary for it to call methods in other objects by using messages. An object sends a message to another object, which receives it, performs some action, and returns a result to the sender.

c) The individual methods are described in text, possibly with one or more lines of pseudo-code. Some typical examples of pseudo-code have been mentioned in chapter 3 in the section "Object-oriented modelling".

Primary result of OOA for the problem domain

The primary results of OOA are a class diagram and the associated CRC cards (as described in chapter 3). The CRC cards are used to describe the individual classes in more detail. Other results include object diagrams and state diagram (these are not discussed further here).

Class diagram

The classes and their structure are drawn in a class diagram. Attributes and methods are filled in on the CRC cards (chapter 3); therefore, the attributes and methods can be omitted from the class diagram. An example of a class diagram is shown in figure 6.8.

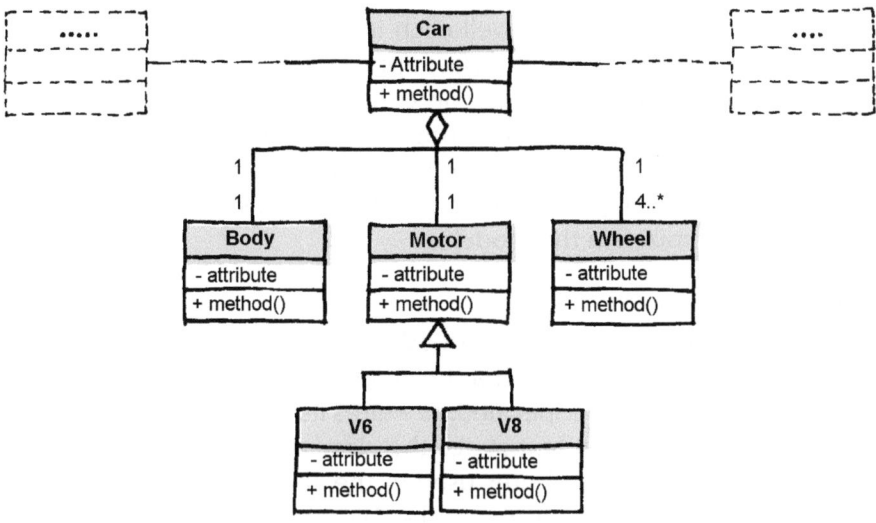

Figure 6.8. *Class diagram.*

The class diagram shows an aggregation structure, which includes the classes Car, Body, Motor and Wheel, together with a generalization structure, which includes the classes Motor, V6 and V8. For the aggregation structure, there is also a specification of cardinality between the classes, for example that Car has 4 Wheels or more.

Flexibility and comprehensibility of the model depend on the coupling between the classes and their coherence. Coupling expresses how closely two classes are connected. Four types of coupling are named below: the first has low coupling and the latter has high coupling [Mathiassen, 1998].

- External coupling: A class refers directly to a second class' public methods.
- Internal coupling: An method in a class refers directly to other, private properties (attributes) in the same class.
- Coupling from below: A class that is a specialization of a second class refers directly to private properties (attributes) in the superclass.
- Coupling from the side: A class that is not a specialization of a second class refers directly to private properties (attributes) in this second class.

A high degree of coupling between two classes means that changes in one of them will lead to changes in the other. It is therefore a negative property, which must be minimized.

Coherence expresses how well each individual class hangs together. High coherence means that the class appears to be a whole, with significant relations between the parts. Low coherence means that the relations between the parts in a class are arbitrary and loose. Attempts to divide classes with high coherence lead to high degrees of coupling, which decreases readability. High coherence is therefore a positive property, which should be maximized in the model.

CRC cards

The individual classes are described by using so-called CRC cards (cf. figure 3.12), as described in chapter 3. On the CRC card, the class' attributes and methods are specified, as well as its relations to other classes; a description in ordinary text states the class' task, and there is possibly a sketch showing the attributes used in the class. In addition, there are areas for describing who has made the CRC card, the date, and possibly the version number.

The application area

The principles of object-oriented analysis described until now have primarily focussed on the structure of the system. Now, we introduce briefly how the system's usage properties can be described, as well as its interaction with the environment, and some basic criteria for producing the system's specification of requirements. The task of producing the requirements specification for a configuration system is also dealt with in more detail in chapter 8.

The system's usage properties fall into three main groups: The system's interaction with actors (which use cases should be available in the system), its functions (the system's content of functions), and its interfaces (user interfaces and system interfaces).

The system's interaction with actors

In this phase, the use cases for the future system are specified. The use cases give a general view of the requirements for the system (focus on the users'/actors' requirements and wishes), and they provide a framework for constructing and evaluating more fundamental requirements for functions and interfaces.

It is important that the system being developed is adapted to the needs of the users from a working, organisational and technical point of view. Thus, it is important to describe the actors who are to use the system, and their desired use cases. An actor is a role that includes users or other systems that have the same use patterns. A use case is a pattern for a limited interaction between a system and actors in the area of application. Note that an actor can also be an IT system, which delivers and fetches information from the system.

For describing actors and use cases, a use case diagram from UML can be used. The notation is briefly illustrated in figure 6.9.

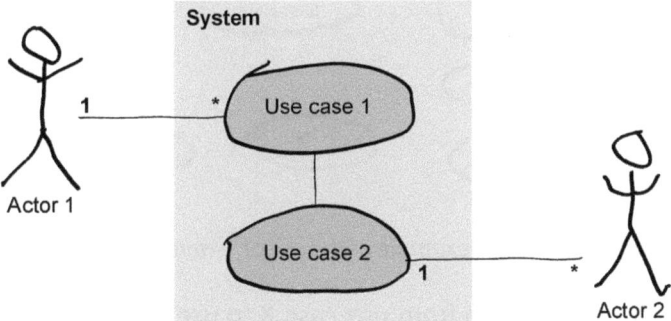

Figure 6.9. *Notation for use case diagram.*

The use case diagram can be used to give a graphic illustration of the actors and use cases, which the future system has to be able to deal with. The use of use cases is an important element in the process of determining the dynamic requirements for a system, i.e. the interaction between the system and the system's future users (actors).

Use case diagrams have 4 main elements: The actors with which the system cooperates, the system itself (the system interface), use cases or functions, which the system can perform, and the lines that represent the relations between the elements [Booch et al., 1999]. This is illustrated in figure 6.10.

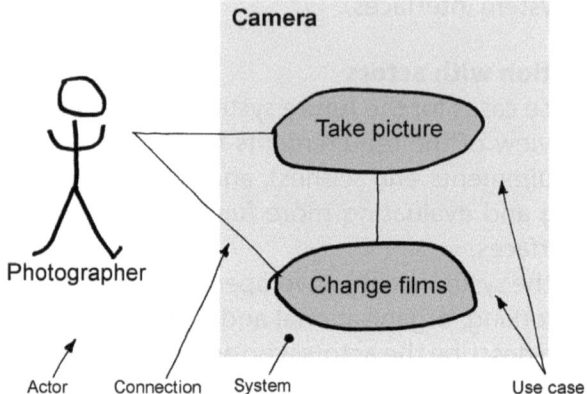

Figure 6.10. *An example of a use case diagram*

To connect use cases, the notations "uses" and "extends" can be used [Booch et al., 1999]. This is illustrated in figure 6.11.

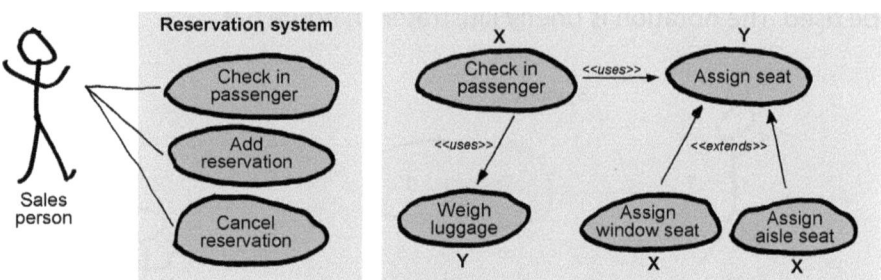

Figure 6.11. *An example of the use of "extends" and "uses"*

A "uses" arrow is drawn from use-case X to use-case Y, to indicate that task X always results in task Y to be performed at least once. "Uses" can be understood as a connection, since it indicates that the use case in question consists of a number of sub-tasks that have to be performed before the entire task is performed.

An "extends"-arrow is drawn from use-cases X to use-case Y to indicate that task X is a special case of the more general task Y. "Extends" is used in situations where the system has a number of use cases which all have the same sub-tasks in common, but where each individual use case includes something special, which makes it inappropriate to collect them into one use case. It is not possible to represent the sequence of tasks in use case diagrams. For this, it is necessary to use other UML diagrams, such as state charts. The reader is referred to Booch et al. [1999] for a more detailed description of use case diagrams.

Functions

The different functions that must be accessible to the actors in the IT system have to be defined: for example searching, registration, updating, optimization, printing etc. It is important to make a complete function list. An incomplete function list will give problems during the programming phase. The sources for the individual functions are partly the description of the problem domain, expressed in terms of classes (attributes, methods) and partly the description of the application area, expressed in terms of its use cases.

Interfaces

Here the user interface and system interface must be described. The user interface is defined as a collection of facilities made accessible to the actors as they use the system. The system interface is defined as a part of a system that implements interaction with other systems and equipment.

When the interaction takes place via a human actor, it is the user interface that is involved, while an interaction via a non-human actor happens via the system interface.

The user interface is, as stated, the part of the system that implements the interaction with the users. The form of dialogue must be defined and some sketches of the desired user interface (screen shots and print-outs) must be worked out.

With respect to the system interface, it must be decided whether the desired technical connections to other systems can be implemented.

Dynamics

If it is necessary to determine parameters in the classes in a particular order (sequential requirement), a need will arise for specifying the program dynamics. The dynamics are often described in a diagram. Figure 6.12 illustrates the door example, where the components have to be determined before the process sequence can be specified, just as the total cost price cannot be determined until both the components and the process sequence have been determined.

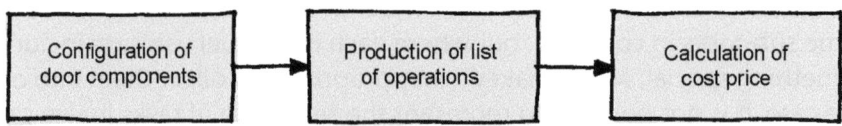

Figure 6.12. *Program dynamics.*

Developing an OOA model for Doors Inc.

We now illustrate Phase 3 of the procedure by developing an OOA model for our example of the specification of doors and their manufacturing process at Doors Inc.

The system is to be used by the door company's sales staff and order processors. Figure 6.13 shows a use case diagram for the application area.

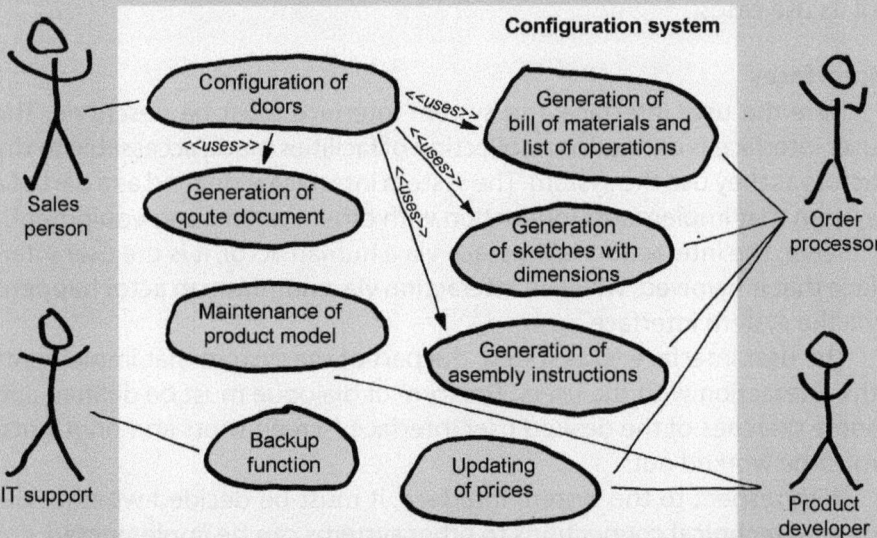

Figure 6.13. *Use case diagram.*

Then, a class diagram is produced, which describes the domain for producing door specifications – quote, bill of materials, sketch with dimensions, list of operations, and assembly instructions. The class diagram can be seen in figure 6.14.

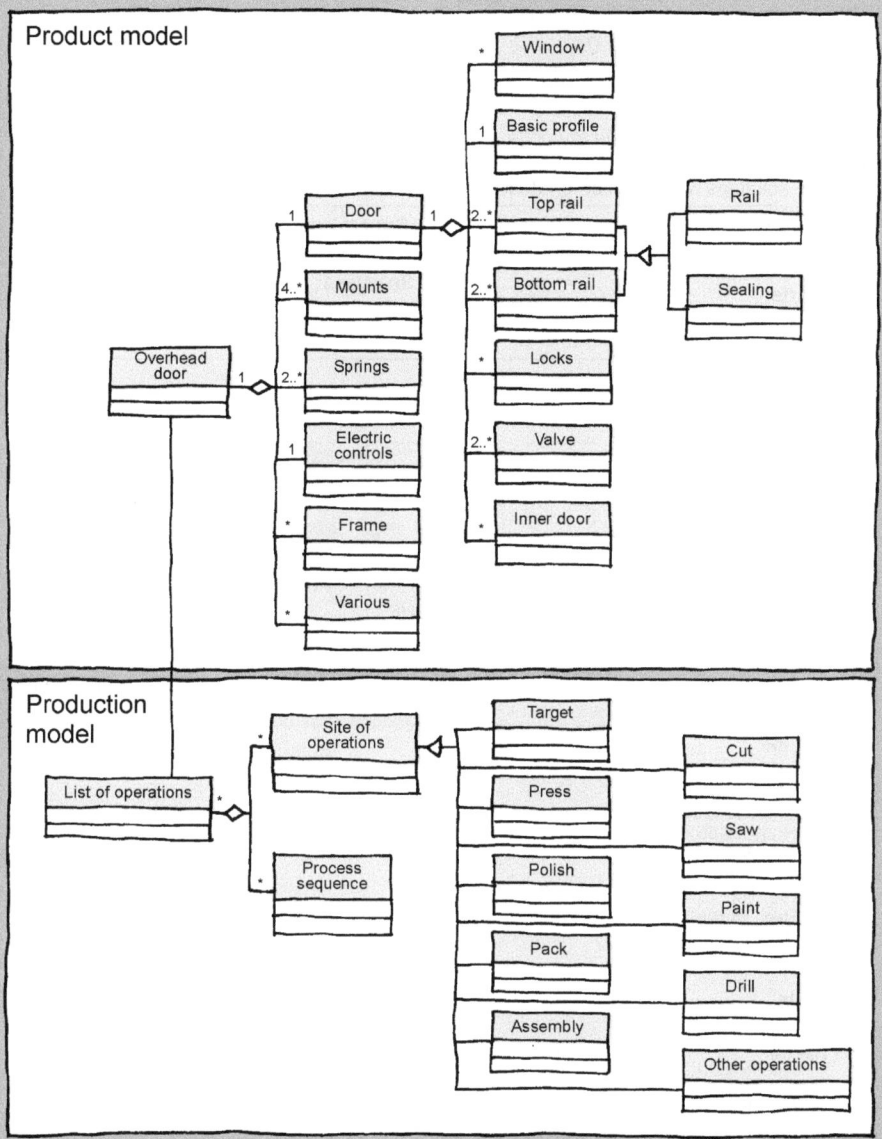

Figure 6.14. *Class diagram for the door example.*

The model involves two subjects, corresponding to the division into a product model and a production model. Classes and class hierarchies are identified, based on the product variant master that has been created. To describe the individual classes, an extended version of the so-called "class-responsibilities-collaborations cards" (CRC-cards) are used. A CRC-card for the class "door" can be seen in figure 6.15.

Class name Door	Revision no. 2.0	Date 01-01-07	Status Implemen- ted	Author KKE

Class purpose:
The class contains knowledge about the door component of overhead doors. The class ensures that only legal configurations of doors are generated. Thus all lists of parts for the door are solutions which can be produced and which are correctly dimensioned.

Superparts: Overhead door	**Superclass:**
Subparts: Window, Basic profile, Top rail, Bottom rail, Locks, Valve, Inner door	**Subclass:**

Sketch

Window

Door height

Door width

Responsibilities	**Class collaborations**
Attributes	
Door type [A, B, C, D] Door width : [[3000-12000] mm Door height: [2000-5000] mm Door slant: [0-45] degrees Sealing: [80-100%] Painting [Yes, No] Material [Steel, Aluminium] No. of windows [0-14] No. of doors [0-4]	
Methods	
Door type [A] Door width < 9 m	
Door width > 7m Material [Steel]	
Material [Aluminium] Painting [No]	
Door type [C] Window shape [round edges]	Window

Figure 6.15. *CRC card for the class door.*

The first step is a division into subjects, which in this case consist of a product model and a production model. Then, the classes and class hierarchies are identified (on the basis of the product variant master already developed), and finally, the individual classes' attributes and methods are defined. In practice, all this takes place in an iterative process, during which the focus area and degree of detail are slowly extended.

Generalization structures and aggregation structures are used to organise the classes. The first type categorize classes with common properties from general to specific objects in the so-called generalization-specialization hierarchies, also known as inheritance hierarchies because the specialized classes further down in the hierarchy "inherit" general properties from the general (superior) classes. In the second type of structures (aggregation structures), classes of different types are collected into whole-part hierarchies, also known as collection hierarchies or composition hierarchies.

In the product variant master for doors and their manufacturing process, some classes can be immediately identified. Overhead door is obviously a class, and since the door consists of several parts (classes), such as the door itself, the fittings, springs, electric control etc., the first hierarchy, a "whole-part" hierarchy, is thus identified, where the overhead door (the whole) consists of various parts.

Other obvious classes are the operational stations the overhead door has to pass through during production. Each of these stations is characterized by the operation to be carried out there, i.e. the description is specific for each station. This relationship is nicely captured in a generalization structure, where the station is categorized according to its process type: saw, drill holes, paint, assemble, insulate etc.

In addition to class hierarchies, the information point of view determines the information (attributes) that describes each individual class (what the class knows about itself). For the overhead door and its components, it is the structural description of an overhead door that is to be represented in the model, i.e. a description of the overhead door's construction and the individual parts in the door.

The individual classes are described by attributes such as: door type, door width, materials etc. The design of the overhead door's components, such as the bottom rail, is constrained by the dimensions specified for the door as a whole. So the geometrical measurements of the bottom rails have to be coordinated (specified in a collaborative way) with the measurements for the overhead door. The other classes correspondingly contain the information (attributes) that are necessary for describing the classes (the classes' knowledge about themselves), within the limitations prescribed by the configuration system.

The functional view of object-oriented analysis focuses on the functionality of the individual classes and the functionality of the system as a whole. This is done by connecting a behaviour with each individual class by identifying methods (services) producing this behaviour, and by describing the collaboration between classes that implement parts of the system's functionality.

Generating a list of operations requires cooperation between several classes. To solve this problem, it is necessary to set up two new classes – site of operations and process sequence. The class "site of operations" organises (contains knowledge about) the operational sites according to the processes executed, while the class "process sequence" contains knowledge about how a process sequence for an item (e.g. "door") must be generated on the basis of the item's characteristics.

In this way, the task of generating lists of operations for "door" is divided into three subtasks. The first subtask is to find the right process sequence. This is done in collaboration with (via a message call to) the list-of-operations class, which, based on information about the door's characteristics and by using its process rules (process sequence) and knowledge of the company's processes (site of operations), can generate a suitable process sequence and return this to "door".

After this, the class "door" is able to consult each individual process in the process sequence and find the most suitable operational site (machine) and, if necessary, alternative operational sites, which can perform the process. In other words, based on information about the door's dimensions given in "door", the process sequence class is able to consult the operational sites that implement the process in question and discover whether the operational sites can work on the door under consideration and can give priorities to possible alternatives, and then return information about the choice of operational site for performing the process.

Once it has made the choice of operational sites (machines), the class "door" is able to consult the individual site of operations classes and ask for the relevant operation time. In this way, the list of operations now comes to consist of a sequence of operations with possible alternative choices of machine, together with a specification of the operation times for the individual operations.

In connection with the OOA analysis, a series of sketchs of the user interface are also produced.

In this case the user interfaces are structured with a series of tabs where the user can write in the requirements for the various parts of the door. There is also a tab to be used to print out offers and manufacturing speci- fications.

The example shown here with the modelling of a system for specify- ing overhead doors and their manufacturing process illustrates the pro- cedure for developing an OOA model. In the following two chapters, we give a more detailed description of how knowledge can be represented in a configuration system (chapter 7), and a number of criteria for select- ing standard software for product configuration (chapter 8).

Bibliography

[Adiga, 1993]: Adiga, S; "Object-Oriented Software for Manufacturing Sys- tems", Chapman & Hall 1993; ISBN 0-412-39750-1.

[Arngrimsson, 1992]: Arngrimsson, Gear; Concepts of Object-Orientation, Proceedings of Sixth IPS Research Seminar, Fuglsø, Denmark, March 1992.

[Bellin & Suchman, 1997]: Bellin, David; Suchman, Simone Susan; The CRC Card Book, Addison-Wesley, 1997.

[Bennet et al., 2002]: Bennett, Simon; McRobb, Steve and Farmer, Ray; Object Oriented Systems Analysis and Design Using UML 2nd ed.; The McGraw-Hill Companies, 2002. ISBN 0-07-709864-1.

[Boehm, 1988]: Boehm, W; A Spiral Model of Software Development and Enhancement, in Thayer, R. H. (Ed.), Los Alamitos, CA, IEEE Computer Society Press, 1988.

[Booch, 1991]: Booch, Grady; Object Oriented Design; Benjamin/Cum- mings Publishing Company, Inc., Californien, 1991.

[Booch et al., 1999]: Booch, Grady; Rumbaugh, James; Jacobson, Ivar; The Unified Modeling Language User Guide, Addison-Wesley, 1999.

[Coad & Yourdon, 1990]: Coad, Peter; Yourdon, Edward; 'Object-Oriented Analysis' ISBN 0-13-629981-4, Prentice Hall, 1990.

[Coad & Yourdon, 1991]: Coad, Peter; Yourdon, Edward; 'Object-Oriented Design' ISBN 0-13-630070-7, Prentice Hall, 1991.

[Cockburn, 1997]: Cockburn, Alistar; Structuring Use Case with Goals, Journal of Object-Oriented Programming, Sep-Oct 1997 & Nov-Dec 1997.

The Unified Modeling Language User Guide, Addison-Wesley, 1999.

[Graham, 1991]: Graham, Ian; Object Oriented Methods; Addison-Wesley Publishing company 1991, ISBN 0-201-56521-8.

[Kruchten, 2000]: Kruchten, Philippe; The Rational Unified Process An Introduction Second Edition, Addison-Wesley, 2000.

[Leffingwell & Widrig, 2000]: Leffingwell Dean, Widrig Don; Managing software requirements – A Unified Approach, Addison-Wesley, 2000.

[Mathiassen et al., 1998]: Mathiassen, Lars; Munk-Madsen, Andreas; Nielsen Peter Axel; Stage, Jan: Object-oriented analysis and design (in Danish), Forlaget Marko, 1998.

Object Management Group. An organisation which inter alia disseminates knowledge about UML via the web site "www.omg.org"

[Rumbaugh, 1991]: Rumbaugh, James; Object-Oriented Modelling and Design, ISBN 0-13-630054-5, Prentice Hall, 1991.

[Warmer & Kleppe, 1999]: Warmer, Jos; Kleppe, Anneke; The Object Constraint Language, Addison-Wesley, 1999.

7

Knowledge Representation and Forms of Reasoning for Expert Systems

In this chapter the basic principles for knowledge representation and forms of reasoning in expert systems are described. A description is also given of the relationship between artificial intelligence, knowledge-based systems and expert systems. Commonly occurring characteristics of expert systems are described in detail, after which the relationships between the representation of knowledge in the knowledge base of expert systems and the control of their reasoning are discussed. The purpose is to illustrate various forms of knowledge representation and reasoning in expert systems for use in developing configuration systems. This is relevant for providing insight into and understanding of which software possibilities exist for programming configuration systems.

Artificial intelligence, knowledge-based systems and expert systems

There is no generally accepted definition of the concept of artificial intelligence. The following quotation is a well-known definition from Østerby [1992]:

"Artificial Intelligence (AI) is the totality of attempts to make and understand machines that are to perform tasks that, until recently, only human beings could perform and to perform them with effectiveness and speed comparable to a human".

Artificial intelligence includes the following areas:

- The ability to make a machine choose a good sequence for some possible actions in order to achieve a particular aim (planning).
- The ability to make machines interpret what they can see (image recognition).
- The ability to make robots move themselves in relation to their surroundings.
- The ability to make machines communicate in natural language (natural language handling).
- Proofs of mathematical theorems.

Another area of artificial intelligence is performing tasks requiring specialized abilities or training. These are usually known as expert tasks. Some examples from Jackson [1992] include:

- Interpretation of data (e.g. signal analysis).
- Diagnosis of functional faults (e.g. faults in equipment or human diseases).
- Structured analyses of complex objects (e.g. chemical components).
- Configuration of complex objects (e.g. a computer).
- Planning of execution sequences (e.g. for robots or sequences of operations for production machines).

Systems which handle these expert tasks are known as expert systems. Jackson's [1999] definition of an expert system is:

"An expert system is a computer program that represents and reasons with knowledge of some specialist subject with a view to solving problems or giving advice".

Østerby [1992] uses the following definition:

"A software system which demonstrates a high degree of performance in the solution of limited problems which would otherwise require considerable human experience".

Typical examples of expert systems are product configurators. These are described by Soininen [1996] as:

"The configurators with the capabilities for checking and producing configurations on the basis of an explicitly represented configuration model are expert systems or knowledge based systems. They automate tasks previously done by human product experts and use the explicit knowledge to reason about product configuration models and configurations".

The term "knowledge-based systems" is often used as a synonym for expert systems, even if the concept is somewhat broader. A knowledge-based system is a system which performs tasks by using "rules of thumb" (pre-defined rules) on a symbolic representation of knowledge (domain knowledge) instead of primarily using algorithms and statistics. This is true of both expert systems and knowledge-based systems. But a knowledge-based system does not need to contain specialist knowledge or expertise within the area to be dealt with, whereas an expert system should do so. Thus, a knowledge-based system that can discuss the weather like ordinary human beings is not an expert system. An expert system in this area should be able to produce weather forecasts etc., i.e. some form of expert knowledge [Jackson, 1999]. Thus, expert systems are a subset of knowledge-based systems, while both of them belong to the area of artificial intelligence (AI).

Characterization of expert systems

Some characteristics of an expert system are, according to Østerby [1992]:

- Reasons from domain-specific knowledge, which may comprise symbols as well as numbers.
- Uses domain-specific methods, which may be heuristic as well as algorithmic.
- Offers good performance within its problem area.

- Explains both what it knows and the reasons for its answers.
- Maintains its flexibility with respect to extensions.

An expert system can therefore be considered an auxiliary tool in a decision situation. This means that any area requiring decisions to be made can in principle be an application area for an expert system.

An expert system can, according to Jackson [1999], be distinguished from a more traditional software system:

- It emulates human reasoning within a problem domain, rather than simulating the problem domain itself. This often distinguishes expert systems from better known software systems that are based on mathematical modelling and computer animation. This does not mean that the system has to emulate the experts' way of thinking completely, but rather that the focus is on emulating the experts' problem- solving methods.

- It performs reasoning on a representation of human knowledge, in contrast to numerical calculations and data retrieval. Knowledge in expert systems is normally expressed in terms of a specially adapted language, where the actual code performing the reasoning is separated from the more general knowledge. The system is divided into what is known as an "inference machine" and a "knowledge base".

- It solves problems primarily by the use of heuristic or "approximate" methods, which are not certain to find a solution to the problem. Heuristics are basically "rules of thumb", which describe how problems in the domain should be solved. This is in contrast to methods that primarily rely on mathematical algorithms, where a result always appears. It is not possible, however, to create algorithms that can solve such complicated problems as expert systems can. Algorithms also play a part in expert systems, but they are not responsible for the primary reasoning.

An expert system can be distinguished from other types of artificial intelligence (AI) [Jackson, 1999]:

- It works with realistic problems which would normally require considerable human expertise. AI systems are often used for research and therefore focus on abstract mathematical problems or simple examples of real problems ("toy-problems") to produce insight or to refine techniques. Expert systems, on the other hand, solve original research problems or problems of commercial interest.

- It typically offers high performance as regards speed and reliability.

This is often not so essential with other forms of AI. Expert systems must be efficient; otherwise they would often not be economically advantageous for supporting the tasks of experts.

- It must be able to explain and verify solutions, as there are often many users of the system, and not all of them are experts in the area concerned. In AI, it is often the developers themselves who use the systems they have developed, and this reduces the need for well-designed and unambiguous interaction (e.g. there are reduced requirements for user interfaces).

Figure 7.1 illustrates the position of expert systems within artificial intelligence.

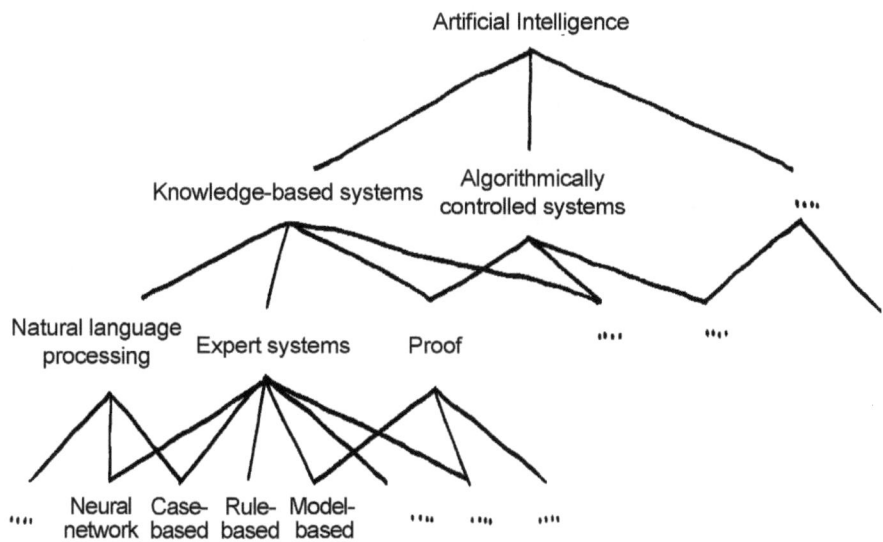

Figure 7.1. *Expert systems*

As previously mentioned, expert systems are a subset of knowledge-based systems, which include natural language processing systems and proof systems. No consensus is found in the literature as to the correct way to make distinctions for expert systems. In this presentation, the classification in Sabin et al. [1998] is found to be the most appropriate. Expert systems are then typically divided according to their forms of reasoning into neural networks, case-based, rule-based and model-based systems. Model-based systems can in turn be divided into constraint-based, logic-based and resource-based [Sabin et al., 1998]. Rule-based, constraint-based, resource-based and case-based types of reasoning are dealt with in more detail later. Logic-based reasoning, however, is not described further in this presentation. For a more detailed description of this form of reasoning, we refer the reader to Sabin et al. [1998]. Neural networks are not described further in this presentation either.

There are basically two forms of reasoning. The first is based on "pattern of direct inference" (deduction machines), i.e. a method in which existing knowledge, represented by well-defined rules, is used as the starting point for generating new knowledge; this includes case-based, rule-based and model-based forms of reasoning. The other form of reasoning is neural networks (induction machines), which take as their starting point a set of information (knowledge), which is then used to deduce rules for generating this information (knowledge) [Roger, 1991].

The first known IT-based configuration systems were expert systems developed by the people who were to use them. In more recent years, however, the tendency has been that many of the systems are developed by the use of standard systems. This development has resulted from technological developments in both hardware and software.

Structure of expert systems

Figure 7.2 shows a typical system architecture for expert systems.

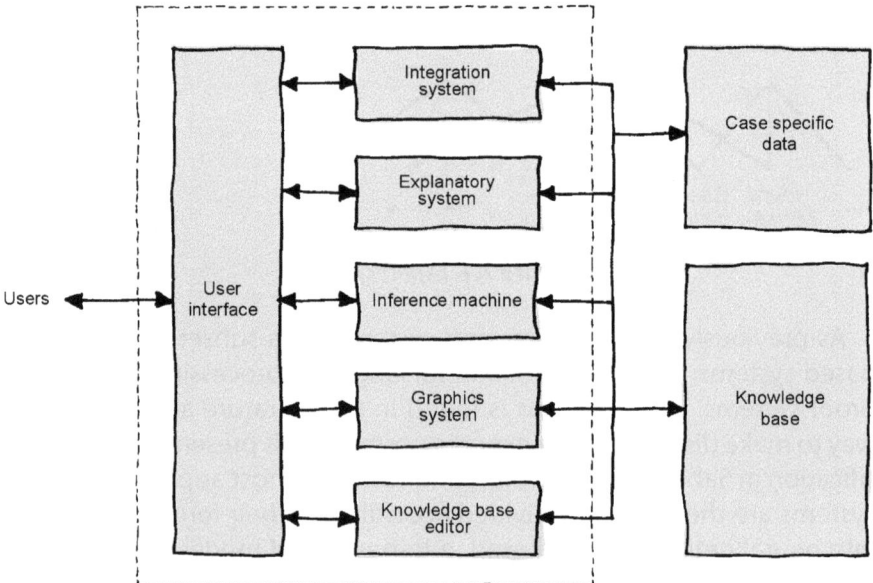

Figure 7.2. A typical structure for expert systems, adapted from Cawsey [1998]

The expert systems (implementation tools) offered by companies selling standard software for configuration can roughly be divided into independent applications, integrated systems (originating from the ERP market) and independent cores. The difference between independent cores and independent applications is that the latter have a more well-developed development environment (including a knowledge base editor, an integration system and a graphics system.

The main individual components in the architecture are briefly described below:

Users of the expert system

Users typically include internal personnel from the company such as those connected with sales, design, support etc., or external customers or suppliers.

The expert system's user interface

All interaction with the users takes place via the user interface. This can be controlled via menus, natural language, graphic interface etc.

The expert system's knowledge base

The knowledge base is used (programmed) to incorporate the knowledge to be included in the expert system (the computer models). There are a number of different ways knowledge can be incorporated into the knowledge base. These are known as the knowledge representation forms for the knowledge base. Because the knowledge base has to work together with the "inference machine" (see below), the knowledge representation must possess a well-defined syntax and semantics that match the inference machine [Cawsey, 1998] [Sowa, 2000].

The inference machine in the expert system

The inference machine solves problems on the basis of the content of the knowledge base. This part of the system can be considered as an interpreter, which works by using the knowledge base and the data inputted by the user via the user interface. The inference machine contains "meta knowledge", which is the higher-level knowledge that controls the derivation of knowledge from the knowledge base [Østerby, 1992]. Cawsey [1998] expresses it in this way: "Then an inference engine (or problem-solving component) is used to reason with both the expert knowledge in the knowledge base and [knowledge] specific to the particular problem being solved".

Case-specific data in the expert system

While the user is using the system, current data from the user interface, the inference machine and the knowledge base are stored; this happens in "case-specific data".

The explanatory system in the expert system

The explanatory system gives the user feedback about the expert system's reasoning: for example, why it chose solution 1 rather than solution 2. It is important that expert systems can give clear explanations via an explanatory system. The reasons for this include:

- The users of the system need to be able to ensure that the system's conclusions are basically correct.
- The domain experts need to be able to see how their knowledge is dealt with, so they can evaluate whether the system works correctly.
- During maintenance, test and extension of the system, programmers need to be able to see which procedures the system calls etc.

It is a question of making the system "transparent", which requires a good deal of focus on making it possible to see how the system works, so the various actors can understand what the system does and why.

Knowledge base editor

The knowledge base editor is the part of the expert system that the programmer uses for creating the knowledge base and making the modifications as maintenance work proceeds. This is often known as the development environment [Østerby, 1992].

Integration system and graphics system

The integration module ensures that the system can be integrated into other IT systems, such as CAD, PDM and ERPsystems.

The graphics system deals with the creation of images and their transfer to the user interface. In recent years, more attention has been paid to the integration and visualization aspects of expert systems. This is primarily due to increased standardization in data exchange, increased bandwidth, and new techniques for handling 2D and 3D graphics.

In the following two sections, various forms of knowledge representation in the knowledge base and forms of reasoning in expert systems are described.

Forms of knowledge representation in the knowledge base

Knowledge representation basically focuses on different ways knowledge can be stored and connected (typically seen from the point of view of human logic). There are many forms of knowledge representation. For example, Jackson [1999], Østerby [1992], Cawsey [1998] and Brewka [1996] mention the following types:

- Natural language (English, Danish, German, etc.)
- Musical notation
- Scientific formulas (chemical, mathematical, physical etc.)
- Graphical notations (functional modeling such as IDEF0, process diagrams, activity diagrams, geometry and UML)
- Structured objects (semantic networks, frames, conceptual graphs, object-oriented technologies)
- Logic (logical statements: propositional, predicate etc.)
- Rules (if-then) etc.

Most of these forms of knowledge representation cannot be handled (interpreted) by a computer. Systems that are solely based on mathematical algorithms can indeed be dealt with by a computer, but it is often not possible to find algorithms that can handle the complex problems found in expert systems. The forms of knowledge representation typically used in expert systems are: structured objects (frames, semantic networks, object-oriented principles), rules (if-then) and logic (predicate, propositional).

Cawsey [1998] lists some requirements for a knowledge representation language:

- Representational strength: the language must be able to represent all the knowledge on which reasoning is to take place (representational coverage).
- Reasoning strength: it must be possible by reasoning to deduce new knowledge from basic knowledge.
- Reasoning efficiency: reasoning should take place in an efficient manner.
- Transparency (accessibility): the represented knowledge is found in a form that makes it easy to understand and modify.
- Clear syntax and semantics: the permitted expressions (terms) and

their meaning (the language's syntax and semantics) ought to be known.

- Naturalness: the language should be reasonably natural and easy to use.

Since no knowledge representation language satisfies all these requirements completely, it is important which knowledge representation language is chosen for solving a specific task. Sometimes "reasoning strength" is the factor that determines whether a language can be used; other times, "naturalness" is more important.

We now look at those forms of knowledge representation most often used in expert systems. These are: structured objects (frames, semantic networks), logic (predicate, propositional) and If-Then rules. The object-oriented modelling representation form, which belongs under structured objects, is not dealt with here.

Structured objects

Within the structured forms of knowledge representation can be found semantic networks, frames and the object-oriented technologies [Cawsey, 1998]. Object-oriented technologies are a further development of semantic networks and frames and therefore rely on their basic elements.

Semantic networks

A semantic network is a form of knowledge representation that focuses on the graphic presentation of objects and relationships within the domain. A semantic network is a collection of objects known as nodes. The various nodes are connected to one another via a series of relations. Both nodes and relations have names or designations. It is possible to develop relatively complex system structures using semantic networks, while continuing to maintain a clear presentation of the system.

An advantage of semantic networks is the great flexibility that can be achieved. The fact that the network is built up of nodes (objects) means that it is easy to add new nodes without the existing network being affected significantly. Thus, it is not necessary to re-structure the entire semantic network in order to add new knowledge [Østerby, 1992].

When knowledge is represented by means of semantic networks, it also becomes possible to translate the information contained in the semantic network into other forms of knowledge representation, such as logic ones. However, this does not mean that translation can take place the other way, from a logic language to a semantic network. This is because a simple semantic network can only contain propositions without variables. This is not the case for logical propositions, which may sometimes also contain variables.

Another phenomenon in connection with semantic networks is inheritance. Inheritance means that nodes (objects) on a lower level can inherit the properties possessed by an object on a higher level. Thus, those objects lying on the second or lower levels have the same properties as the object on the first level from which they inherit the properties. The use of inheritance in connection with semantic networks leads to great flexibility. The way this flexibility is achieved is that when something has to be corrected, added or deleted, this only has to be done in one place, namely in the object at the highest level from which the underlying objects inherit their properties. Thus, the same information does not need to be repeated again and again.

Frames

Frames are another commonly used method for representing knowledge. They provide a method for representing facts and relationships. A frame is a description of structured objects with associated descriptions. The various objects contain slots, containing blank fields in which a particular type of information can be written. The description specifies which data types and associated values are valid for the individual fields. This is the beginning of the creation of what we call "classes" in the object-oriented technologies

The object's various fields can be assigned two different types of values [Østerby, 1992]:

- Standard values (defaults)
- Variable values

Standard values are given in advance, for example when configuring a car. When defining the cars features, it is unnecessary to specify the number of wheels, since it is typically four (disregarding the spare wheel). Another case could be when the default value for the number of car doors is set to four, because experience shows this is the preferred number. In this case, however, there might be some purchasers who want a car with only two doors; then it would be necessary to change the pre-defined default value to 2.

Variable values are those about which it is not possible to say anything definite in advance. An example is selection of the car's engine capacity.

You can build a structure in which some objects perform particular tasks - for example, to fetch prices from different objects and calculate a total price - it is possible to collect properties in a single object instead of having the same property spread out over many different objects. This is an advantage when systems are to be maintained. Instead of correcting the same property in many different objects, the property concerned can be corrected in a single place. In this way the probability of errors is reduced considerably. In principle, frames are a form of semantic network.

In frame structures, objects with the same properties and structure can be grouped in classes. It is for example not necessary to have 5.2 million objects used separately to keep track of the name, address, telephone number etc. of each member of the Danish population. Instead, the objects can be organised in a class, whereby they are reduced to a list containing all this information for the whole population. These classes can be collected together into new higher-level classes, so that even more information is collected in fewer but more complex classes.

Logic

Logical derivation takes place when a correct conclusion is drawn on the basis of one or more premises that are true [Cawsey, 1998]. An example could be:

Proposition 1:
It is true that all mammals give birth to living young.

And

Proposition 2:
It is true that the whale is a mammal.

From this one can deduce that whales give birth to living young, even if this information has never been presented previously.

In order to build up a logical knowledge representation, it is essential to set up rules for which conclusions can be drawn in various situations. However, there are a number of different ways of presenting knowledge using logic statements. These include propositional logic and predicate logic.

When product models are expressed in terms of propositional logic, a series of propositions are used, i.e. statements that can be either true or false. For example:

- Denmark is in USA
- Århus is north of Esbjerg
- The lion is a bird
- The Earth is round

Each of the statements in the example is a proposition with a truth value that is either true or false. Propositions can be composed using logical operators (Boolean operators). Propositions containing such operators are known as composite propositions. The operators include:

AND	logical symbol ∩ (disjunction)
OR	logical symbol ∪ (conjunction)
NOT	logical symbol ~ (negation)
IMPLIES	logical symbol → (implication)

It is possible to predict a composite proposition's truth value if the truth values of the individual propositions are known. It is not necessary to know what the various propositions refer to. An example of how the truth value of a composite proposition can be determined is: *"If a proposition X is true and another proposition Y is false, then the composite proposition X AND Y will be false. If the composite proposition is written as X OR Y, then the composite proposition is true".*

Another branch of logic is predicate logic. In predicate logic, it is not enough to consider a proposition as a whole but also the components comprising the proposition. In predicate logic, the predicate itself is the basic element. Predicate is a linguistic term for properties or relationships. Some examples of predicates are shown below:

Predicate		
Is_big (X)	=>	X is big
Loves (X,Y)	=>	X loves Y
Is_human (X)	=>	X is human
Sleeps (Z)	=>	Z sleeps
Gives (U,V,T)	=>	U gives V to T

The predicates have a symbol that specifies properties or relationships, and one or more arguments, which specify who or what the predicate applies to. Schematically, this has the form: Symbol (argument 1, argument 2, ⋯, argument n).

Some predicates can have several arguments and are therefore divided up into one-, two-, three-place predicates etc., depending on the number of arguments. For example:

Predicate	Argument(s)		
Is_big (X)	Ball	=>	The ball is big
Loves (X,Y)	Pia, Peter	=>	Pia loves Peter
Is_human (X)	Peter	=>	Peter is human
Sleeps (Z)	Dog	=>	The dog sleeps
Gives (U,V,Q)	Peter, ball, Pia	=>	Peter gives the ball to Pia

Thus "sleeps" is a one-place predicate, "loves" a two-place predicate, "gives" a three-place predicate.

To move from a predicate to a statement, the blanks, or argument positions, are filled in. By filling in the variable positions, it becomes possible to ask whether it is true that the dog sleeps, that Pia loves Peter etc. On the other hand, it makes no sense to ask about the truth value of an expression where the variable places are not filled in.

A base containing information about a random family could for example be as follows:

- Is_baker (Fred)
- Is_teacher (Annie)
- Is_father_of (Fred, Betty)
- Is_married_to (Fred, Annie)
- Is_married_to (Annie, Fred)
- Is_schoolchild (Betty)
- Has (Betty, Tweetypie)
- Is_canary (Tweetypie)

Once the knowledge base has been created, it can be questioned: Who is married to whom? What is Fred? Is Annie a baker? Predicate logic is especially useful when creating a data base or a knowledge base containing facts that are related.

In general, logic deals with logical statements, i.e. things that are true. Most forms of semantic networks, frames and rules can all be re-written into logic statements without losing knowledge. Moving in the opposite direction usually causes loss of knowledge.

If-then rules

Instead of representing knowledge in relatively declarative, static form (such as a set of relationships that are true), knowledge can be represented as a set of rules, which tell us what is to be done or what can be concluded in different situations. The rules consist for example of statements with the form: If <condition> then <conclusion> [Cawsey, 1998].

In the next section, we discuss forms of reasoning in expert systems.

Reasoning in the expert system

Expert systems for supporting specification processes typically use the rule-based, model-based (constraint- and resource-based) and case-based reasoning forms. This section describes these forms of reasoning. Emphasis is on the rule-based and constraint-based forms of reasoning, as they have most relevance for the construction of configuration systems. The remaining forms of reasoning are only dealt with superficially.

Rule-based

Rule-based systems consist of a database, a rule base, and an inference engine. The inference machine contains procedures that activate the rules in the database in a given sequence and thus generate new information [Cawsey, 1998]. The rules consist of if-then statements, which can be modelled in a decision tree [Faltings et al., 1994]. Searching for and activation of rules takes place either forwards or backwards in the tree structure according to procedures defined in the inference machine. Figure 7.3 illustrates the elements in a rule-based system.

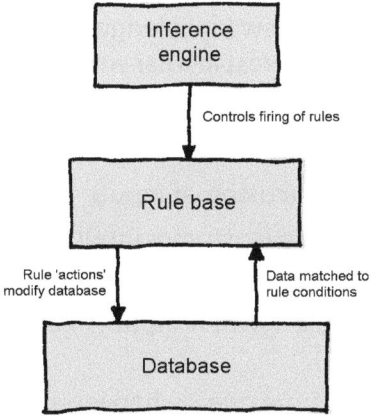

Figure 7.3. *Elements in a rule-based system [Roger, 1991].*

Faltings et al. [1994] illustrate the basic principles in a rule-based system with the following example based on cars:

A car consists of five modules with associated attributes:

- Package type: Deluxe, Luxus, Standard
- Car type: Sports car, Family car
- Motor: A (2.0 litre), B (3.0 litre)
- Gearbox: Manual, Automatic, Semi-automatic
- Body: Convertible, "Hatchback", Sedan

The following rules are used in the example:

- Rule 1: IF Package type = Deluxe and Body = Convertible THEN Motor A
- Rule 2: IF Package type = Deluxe and Body = "Hatchback" THEN Motor B
- Rule 3: IF Package type = Standard and Body = Convertible THEN Motor A
- Rule 4: IF Motor = A THEN Gearbox = Manual
- Rule 5: IF Motor = B THEN Gearbox = Automatic
- Rule 6: IF Car type = Sports car THEN Body = Convertible
- Rule 7: IF Car type = Family car THEN Body = Sedan
- Rule 8: IF Car type = Sports car THEN Gearbox = Manual

By searching forward [Faltings et al., 1994] (rules 1 to 8), the inference machine selects the rules where the pre-conditions are fulfilled and "fires" them in order to conclude new knowledge (facts).

A customer expresses interest in a car with "Package type = Deluxe and Car type = Sports car". The configuration in the rule-based expert system arrives at the following:

- Car type = Sports car (customer input)
- Package type = Deluxe (customer input)
- Body = Convertible (from rule 6 given input Car type = Sports car)
- Motor = A (from rule 1 when input is Package type = Deluxe and Body = Convertible)
- Gearbox = Manual (from rule 8 when input is Car type = Sports car or via rule 4 when input (generated facts) is Motor = A)

If a new car type is introduced, for example a Holiday car, then this will require the following changes in the knowledge base:

- Rule 9: IF Car type = Holiday car THEN Body = Convertible
- Rule 10: IF Car type = Holiday car THEN Gearbox = Semi-automatic

If a customer now orders a "Car type = Holiday car, Package type = Deluxe", then the following rules will be used in a forward search:

- Step 1: IF Car type = Holiday car THEN Body = Convertible (via rule 9)
- Step 2: IF Package type = Deluxe and Body = Convertible THEN Motor A (via rule 1)
- Step 3: IF Motor A THEN Gearbox = Manual (via rule 4)
- Step 4: IF Holiday car THEN Gearbox = Semi-automatic (via rule 10)
- Conflict between Step 3 and Step 4.

The introduction of new rules into the knowledge base can result in existing rules no longer being valid. In the example above, rule 4 is now no longer valid, since it is no longer correct that all cars with motor A must have a manual gearbox. It is therefore necessary to make the rule context-dependent:

- Step 1: Delete rule 4
- Step 2: Add rule 4a: IF Motor = A and Car type = Holiday car THEN Gearbox = Semi-automatic
- Step 3: IF Motor = A and Car type = Sports car or Family car THEN Gearbox = Manual

Maintenance is thus one of the big problems in rule-based systems. Even small additions to the knowledge base can lead to large changes in the existing rules, since the rules are context-dependent.

Constraint-based

Knowledge can be represented as equations or more generally as constraints. For example $v = s/t$, which can be used to calculate v from s and t, or alternatively, t can be calculated from v and s, or s from v and t. This could be an example of a constraint in a system.

Constraints are declarative in contrast to rules, which are procedural.

Constraints in a discrete domain can be expressed as relations between variables (attributes, parts), where the constraints declare which combinations are permitted or not permitted. Faltings et al. [1994] describe a constraint as follows:

> "A constraint describes a relation between components (variables) and all the allowed combinations of values for the variables in the relation. All variable-value combinations, which are not explicitly mentioned, are assumed to be not allowed. The goal of the constraint satisfaction process is to find for each variable values satisfying all the constraints or to realize that there exists no solution for the constraint satisfaction problem".

According to Schwarze [1996], rules and constraints can be described as follows:

- "A rule specifies an 'as-it-is' view of a part of the system
- "Constraints describe the world in terms of logically consistent states."

Thus, constraints and rules have different interpretations. In the following example (based on the example in the previous section), the basic elements of constraint-based reasoning are illustrated:

- Constraint 1: Package type Body Motor g permitted value combinations: (Deluxe convertible A)(Deluxe "Hatchback" B)(....)
- Constraint 2: Motor Gearbox g permitted value combinations: (A manual)(B Automatic)(A Semi-automatic)(....)
- Constraint 3: Car type Body g permitted value combinations:
- (Sports car Convertible)(Family car Sedan)(....)
- Constraint 4: Car type Gearbox g permitted value combinations: (Sports car Manual)(Family car Semi-automatic)(....)

Marketing decides (as in the case of the rule-based system) to introduce a new Car type: Holiday car. This requires the following changes to the knowledge base:

- Constraint 3': Car type Body g permitted value combinations: (Holiday car Convertible)
- Constraint 4': Car type Gearbox g permitted value combinations: (Holiday car Semi-automatic)

In this example, constraint 2 means "Motor type A is possible with a manual gearbox", while rule 4 (in the previous section on rule-based systems) means "every car with Motor type A must have a manual gearbox". Thus, constraint-based systems are context-independent.

Since constraint-based systems are constructed to be context-independent, constraint 2 is for example still valid even if the same changes are introduced as in the case of rule-based reasoning. This means that knowledge from various sources can be integrated into a constraint-based system without severe modifications to the existing knowledge in the knowledge base. This leads to easier maintenance and continued development than in the previously mentioned rule-based systems.

Resource-based

In resource-based reasoning, a "producer-consumer" model is set up [Sabin et al., 1998]. An example is shown in figure 7.4.

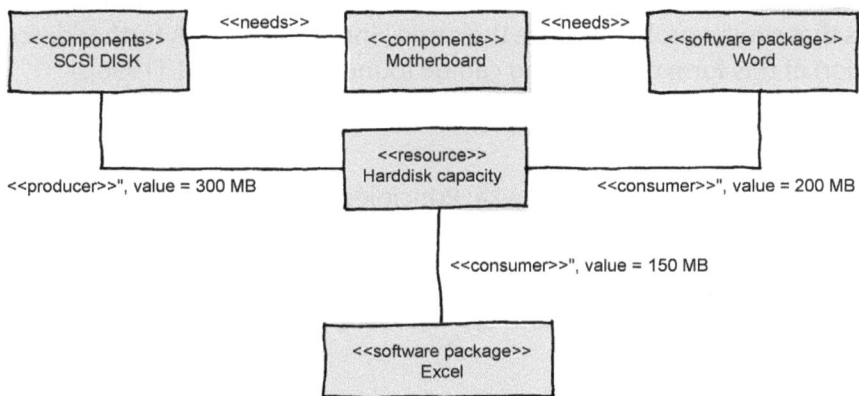

Figure 7.4. *"Producer – consumer" model*

The objects (classes) in the model are assigned a resource, for example to produce 200 units or consume 200 units. Reasoning takes place when the objects are selected or deselected, depending on the state of the resources. If a SCSI disk is chosen (cf. figure 7.4), 300 MB of hard disk space are produced in a resource. A SCSI disk cannot be chosen without a Motherboard being chosen as well. If a software package (Word) is then selected, 200 MB of the 300 MB hard disk space are consumed, which means that there are still 100 MB of hard disk space left; this means that the model is still consistent, since the quantity of produced resources has not been exceeded [Juengst et al., 1998]. If an additional software package of type Excel is chosen, which consumes an additional 150 MB hard disk space, then the quantity of produced resources is exceeded, and it becomes necessary to add yet another SCSI disk in order for the configuration to be possible. A further description of this form of reasoning can be found in Sabin et al. [1998].

Case-based

Case-based (reasoning) systems use previous design cases or configuration cases to solve new problems. In Schwarze [1996], a case is described as follows:

"A case is a contextualized piece of knowledge representing an experience that teaches a lesson fundamental to achieving the goals of the reasoner."

The idea behind case-based reasoning is to copy the way in which experts often solve problems: They make rules on the basis of experience and use these rules in analogous ways for solving other problems, in order to arrive at new knowledge. A precondition for case-based reasoning is that similar problems have the same kind of solution. A further description of this form of reasoning can be found in Sabin et al. [1998].

Bibliography

[Brewka, 1996]: Gerhard Brewka; Principles of knowledge representation, CSLI Publications, Center for Study of Language and Information, Leland Stanford Junior University, 1996.

[Cawsey, 1998]: Alison Cawsey; The Essence of Artificial Intelligence, Printice Hall, 1998.

[Faltings et al., 1994]: Boi Faltings, Rainer Weigel; Constraint-based knowledge representation for configuration systems, Technical Report No. TR-94/59, Departement D'Informatique, August 1994.

[Jackson, 1999]: Peter Jackson; Introduction to expert systems, third edition, Addison-Wesley, 1999.

[Juengst et al., 1998]: Werner E. Juengst, Michael Heinrich; Using Resource Balancing to Configure Modular Systems, Daimler Benz Research and Technology, IEEE Intelligent systems, 1998.

[Roger, 1991]: Kerr Roger; Knowledge-Based Manufacturing Management, University of New South Wales, Australia, 1991.

[Sabin et al., 1998]: Daniel Sabin, Rainer Weigel; Product Configuration

Frameworks – A Survey, University of New Hampshire and Swiss Institute of Technology, IEEE Intelligent systems, 1998.

[Schwarze, 1996]: Stephan Schwarze; Configuration of Multiple-variant Products, BWI, Zürich, 1996.

[Soininen, 1996]: Timo Soininen; Product configuration knowledge: case study and general model, Faculty of Information Processing Science, Helsinki Universitet for Teknologi, 1996.

[Sowa, 2000]: John F. Sowa; Knowledge Representation – logical, philosophical and computational Foundations, Brooks/Cole, 2000.

[Østerby, 1992]: Tom Østerby; Artificial Intelligence – methods and systems (in Danish), Polyteknisk Forlag, 1992.

8

Choosing Configuration Software

In this chapter we describe some factors which influence the choice of software for product configuration. We start by describing the concept of "standard software" (or standard systems), and a series of advantages and disadvantages associated with the use of standard solutions are discussed. Then, various forms of programming for use in programming product models are presented, and a series of criteria for selecting software for product configuration are given.

What is a standard system?

Many different definitions of "standard systems" have been proposed in the course of time. In this presentation we, make use of the definition given by Rosenbeck [1995], who defines standard systems as follows:

"A standard system is an IT system for solving a number of general tasks. A standard system is marketed, documented and maintained by an external supplier for a broad group of companies who have a common need, and it is therefore characteristic that there are relatively many users of the same standard system"

According to this definition, a piece of software is called a "standard system" when there is an external supplier who markets, documents and maintains the configuration software for a large group of customers.

In ESPRIT-AMICE [1989], the CIM-OSA Reference model was developed. This reference model divides IT systems into three basic types: *generic*, *partial* and *particular*.

Generic systems suit all companies. They are standard systems which have a wide range of application, such as operating systems and word processing systems, and are used everywhere in industry. Such systems are relatively cheap, since their development costs are divided among a large number of users.

Advantage	Description
Rapid installation	The standard system has a short delivery time, and thus the project time will be reduced and initial use can take place earlier.
Development & maintenance	Programming of the configuration system will be cheaper, as it can be performed by your own domain experts, and there is no need to use system developers to the same extent as if you develop the system yourself. Most maintenance can also be done by domain experts with support from the supplier of the standard system.
Supplier support	There is the possibility of support from experts during development/operation of the system, if problems arise.
Certain project cost calculation	A relatively certain calculation of costs can be made. This assumes, however, that the system does not need too much adaption to match other systems when installed.
Built-in experience	Standard systems often incorporate a considerable amount of implicit knowledge and competence. The systems are "field-proven" and built up according to "Best Practice". This happens as the suppliers and developers of the standard system accumulate experience from previous and current installations. This ensures that the systems are continually developed, at the same time as new versions appear on the market.
Flexible system	Standard systems are flexible systems, which can be adapted to the changes which companies are exposed to as time goes by. This flexibility ensures a longer life for the system.
Higher system and training quality	As standard systems are flexible and more tested than self-developed systems, they must honestly be said to be of higher quality. That the system is better tested is an advantage in the start-up phase, as this is the critical period seen from an operational point of view. The systems are also better developed as regards documentation and safety routines. As the systems are completely documented, training can be prepared for and carried out on a better foundation than with self-developed systems.
Test before purchase	Standard systems can be tested in various environments before purchase. The company can also build up prototypes to support the decision to introduce the system, and it is furthermore possible to obtain user references for the system.

Figure 8.1. *Advantages of standard systems*

Partial systems are standard systems which are intended specifically for a particular area of industry, such as CAD systems which are used in the ship building industry. As a consequence of their area of use, such systems have a relatively small number of users and are often relatively expensive, depending on the size of the system and its complexity.

The final type, *particular* systems, are company-specific. This means that the system is developed in the company to suit the specific company's needs and requirements. Since the development costs have to be covered by the company, the price of the system is directly proportional to the number of hours of development time and the complexity of the system. Only generic and partial systems can be considered "standard systems".

Knutsson et al. [1997] explain some of the advantages of using standard systems as opposed to self-developed systems. These advantages are presented in figure 8.1.

There are unfortunately not only advantages in using standard systems. Knutsson et al. [1997] also mention a series of pitfalls, as shown in figure 8.2 below.

Disadvantage	Description
Over-hasty decision	It can be dangerous to introduce a standard system on the basis of too poor an analysis or via an over-hasty decision, as the company can be subsequently run into some of the problems mentioned below in this figure.
Supplier dependency	It is not possible to change supplier without considerable costs. The company gets more or less involved with the supplier of the system. If the supplier goes broke or becomes unstable, it may be difficult to get support for the system or to get the system updated as a result of changes in the market and/or the organisation.
Integration/ Adaption	It is important to ensure that the standard system can be adapted to the company's wishes and needs, and can be integrated into the company's other business process and IT systems. Solution of any integration problems may require many resources, since the changes must be carried out by the system supplier.
Altered working conditions	Some users will need to adapt their working procedures to the standard system. This may lead to some of the procedures becoming less efficient.

Figure 8.2. *Disadvantages of standard systems*

A central issue when developing product models is whether the company should develop the system itself or purchase a standard system. The development currently moves towards greater use of standard configuration systems for creating product models, which is also due to the advantages mentioned in figure 8.1. If a standard system is decided upon, it is necessary to find the most suitable software for the problem in hand, and this is often a difficult task.

No software system is best in all situations. Which software is best is situation-dependent; therefore, it is necessary to consider the current situation and then choose the optimal software on this basis.

An overview of forms of programming

Figure 8.3 gives an overview of a number of the most common types of software for programming product models.

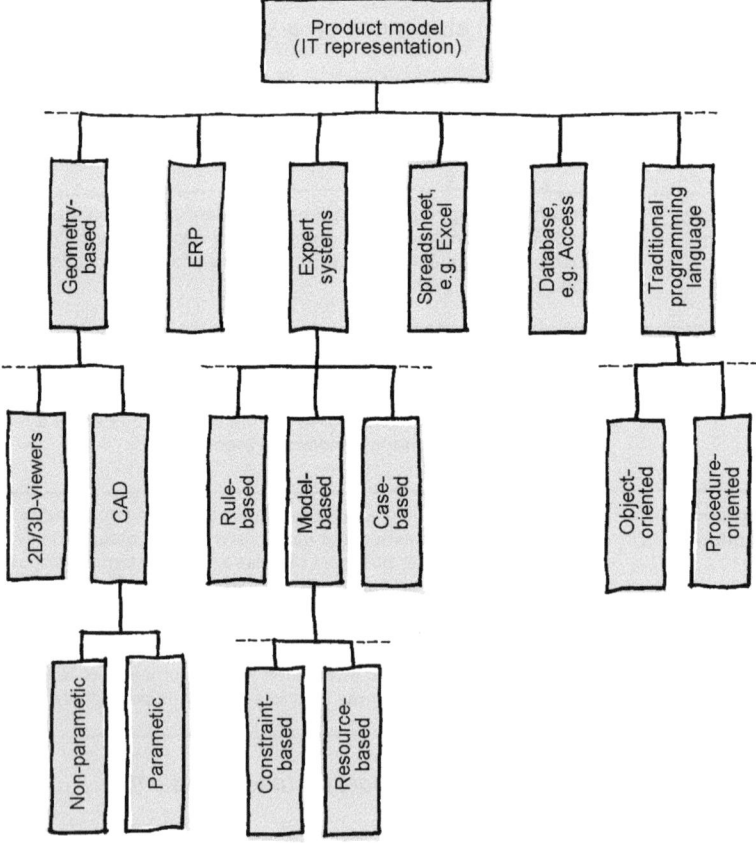

Figure 8.3. *Different types of software for programming product models*

Software solutions are often a mixture of the types of programming shown above. For example, constraint-based and case-based expert systems can be combined, or an ERP system can incorporate a rule-based expert system. The different types of software are explained below.

Expert systems

Expert systems are often used to solve complex configuration problems. They are used for example by F.L. Smidth A/S to configure cement factories. As mentioned above, expert systems can be divided into three basic types: rule-based, model-based and case-based expert systems. Model-based systems can in turn be divided into constraint-based and resource-based expert systems. For configuration of industrial products, rule-based or constraint-based expert systems are commonly used. Experience shows that rule-based expert systems can be difficult to maintain, so most development today is focused on constraint-based systems. All things considered, compared to the other types of programming, expert systems are characterized by high representational strength, high reasoning strength, and high reasoning efficiency.

There are a number of suppliers who sell expert systems for configuration purposes. Figure 8.4 shows some of the best-known expert systems. Several of these systems are subsets of other systems, for example parts of an ERP system.

Standard systems for configuration		
Movex Configurator	INFOR Configurator	Engineering Intent
Oracle Configurator	Trilogy	eLogia
SAP Configurator	3DFacto	E-con
Array Technology	Tacton	Socrates, CinCom
FirePond	ConfigIT	Siebel (eAdvisor)

Figure 8.4. *Standard systems for product configuration.*

Spreadsheets

It is a widespread practice to programme product models in spreadsheets (e.g. Excel). Spreadsheets have proved to be well-suited to simple configuration problems, but if the problems become more complex, then spreadsheets are often not the right type of software to choose. Compared to expert systems, spreadsheets have low representational strength, low reasoning strength and low reasoning efficiency.

Database systems

Database systems (e.g. Access) were basically developed for the storage and retrieval of data, and they are therefore an important element in most expert systems, ERP systems etc. Examples have shown that it is often possible to use database systems for simple configuration. However, database systems often have low representational and reasoning strength and low reasoning efficiency in comparison with expert systems. For that reason, they are not designed to handle complex dynamic configuration without being supplemented by an expert system.

ERP systems

Most ERP systems have a built-in expert system. Often, this is a simple rule-based system closely associated with the structures that describe lists of parts in the ERP system. Thus, the expert system often does not have quite the same functionality as in "independent applications". Oracle Configurator is an example of an ERP system where a constraint-based expert system has been purchased and integrated into the ERP portfolio.

CAD systems

CAD systems typically focus on geometric instances of products (including parametric variation). This means that most CAD systems cannot handle real product configuration for creating product variants. It is therefore becoming more and more common that for example standard configuration systems (expert systems) are integrated into CAD systems. The configuration system deals with finding the legal choices (including the choice of components), while the CAD system draws the corresponding products.

2D/3D viewers

2D/3D software solutions (outside the CAD system) are available which can produce a visual representation of the configuration. Such systems are often connected with traditional expert systems, where the expert systems handle the logical bindings (e.g. the limitations on components), while the 2D/3D software solutions deal with the geometrical bindings and placement in 2D or 3D. There are two basic types of visual configuration: either to start with a text-based configuration, which is then displayed visually, or to perform the configuration directly in the visual environment.

Traditional programming languages

Self-developed systems are often built using a programming language such as Java, C++ or PowerBuilder. Basically, all configuration programmes can be self-developed, but it often requires enormous resources to develop the same functionality as is found in standard systems. It is difficult to incorporate high representational and reasoning strength and high reasoning efficiency in a self-developed application. Likewise, it requires considerable resources to develop a well-functioning knowledge base editor, a well-functioning integration system, and a well-functioning explanatory system.

Criteria for choosing standard software for product configuration

In this section, we describe a number of criteria relevant when selecting standard software for configuration. The criteria are divided into "criteria for choosing software" and "criteria for choosing software supplier". The criteria are partly based on Buch [1999].

"Criteria for choosing software" are classified under a series of descriptive characteristics for the IT system based on a system point of view in relation to the following:

1. Software functionality
2. Software structure
3. Software interfaces (external relations to the system, including system integration)

In the software selection process, it can be convenient to distinguish between primary and secondary criteria. The primary criteria are critical for whether the project can be carried out, while the secondary ones are "nice to have". An efficient way of getting some insight into suppliers' standard configuration software is to give them a task (possibly a fictive one), which they are to solve before they present their software.

Criteria for choosing software

Software functionality

Dynamic configuration

A software system deals with dynamic configuration, if the given solution space is reduced for each individual choice that is made through the configuration, and if the system also helps the user revise a choice. Thus, if the user of the system wishes to change a previously made choice, then the system indicates what has to be altered so that the change can be made.

Price- and cost calculation (support for pricing principles)

Can the configurator deal with complex pricing principles for price and cost calculations?

1. How are prices maintained?
2. How are various prices handled on different markets?
3. How are the various monetary standards handled?
4. How are the various units (kg, pounds etc.) handled?
5. How are customer specific discounts etc. handled?

Online/offline configuration

If the system is based on a server, it is an advantage if it is possible to perform a configuration both on-line and off-line with the server.

Report/quote generation

In connection, for example, with the generation of quotes, the configurator must, as a standard feature, be able to generate reports as documentation for the concrete configuration (input parameters, choice of product, price calculation etc.). Generation of quotes typically takes place in Word or HTML.

Dealing with sub-models

It can often be an advantage if the software can handle sub-models. This makes it possible for example for several developers to work on the individual sub-models at the same time. It also facilitates maintenance.

Version control of product models (synchronization)

It ought to be possible to control the product models so that only valid product models are accessable. It may also be necessary to deal with previous models (configurations), due for example to repairs or re-acquisition of spare parts.

Version control of software

The company should be able to control and manage the current versions of the software.

Backup

The company should be able to perform automatic backup procedures for both product models and software.

Administration of users and system

The system must possess functionality to perform the necessary administrative tasks associated with its operation of a distributed environment – such as establishing and maintaining users, and controlling current user privileges. Another important administrative task is opdating users' models. Is this done by using the Internet, cd etc.?

Software structure

Type of standard system

Which type of standard system is being considered? Is it an independent application, an integrated module in an ERP system (an integrated system) or an independent core? An independent core primarily consists of an inference machine (an evaluation system) and knowledge base (cf. figure 7.2), whereas an independent application basically also contains the surrounding sub-systems (cf. figure 7.2); e.g. it may also contain a knowledge base editor, explanatory system, graphics system etc.

Use of technology – client/server environment

Can the system support the IT infrastructure used by the company, for example a client/server structure and a Windows platform?

Adaptation of standard systems

In connection with adaptation of the standard system, it should be possible to use known development tools such as C, C++, VB and Java.

Web-enabling

Can the above-mentioned functionality requirements be supported through the use of web technology (web servers, web browsers etc.)?

Inference machine (control structure)

The configurator must be built up around an efficient and well-structured inference machine, which can be considered as interpreter or control structure for the rules modelled in the system. The inference machine "determines" which heuristic search technique is to be used in order to decide how the knowledge in the system's knowledge base is to be applied to the problem in hand, i.e. which form of reasoning does the software use? Is it for example a rule-based, model-based (including constraint-based and resource-based) or case-based form of reasoning that is performed?

Time of reasoning

At what stage does the standard system perform reasoning? Is it during compilation, at runtime before making any choices in the configurator, or at runtime after the necessary choices have been made in the configurator? If the configurator is to operate on the Internet, it is often a good idea to perform reasoning during compilation, as this usually leads to short response times.

The form of representation in the knowledge base

Which form of representation can be used in the knowledge base? Initially, the following questions can be asked:

1. Which form of knowledge representation can be used in the knowledge base to represent, for example, classes and attributes?

2. Which form of knowledge representation can be used in the knowledge base to represent methods? Is it for example if-then rules, Boolean expressions, or arithmetic expressions? One of the most common forms of representation for methods (rules) in constraint-based expert systems is to use Boolean and arithmetic expressions. Boolean expressions make use of logical operators such as AND, OR, NOT, TRUE, FALSE, while arithmetic expressions use operators such as +, -, *, /.

3. Is the knowledge base based on fundamental object-oriented principles? For example: the object-class relation, instances, inheritance, messages, encapsulation and abstraction.

4. Can tables be used in the knowledge base? The use of tables to represent methods can be extremely efficient. An example of an expert system using representation by tables in the knowledge base is shown in figure 8.5. The table in figure 8.5 illustrates some methods which put limitations on the relationship between a product's

materials (such as leather or steel) and the quality of the product (such as discount or deluxe). If (as in the figure) a watch of type "discount" is chosen, then its strap can only be made of leather. If the type "deluxe" is chosen, then it can be obtained in both steel and gold. Tables are a relatively strong form of representation, and they are easy to understand.

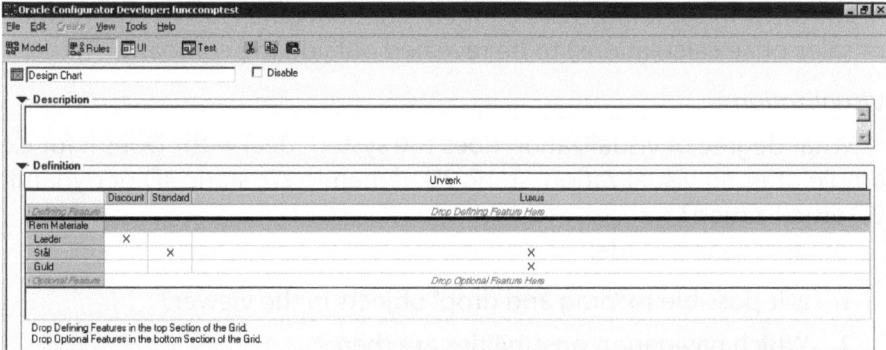

Figure 8.5. *Example of methods represented via tables. Source: Oracle Corporation, all rights reserved.*

5. Does the system provide a syntax check of the knowledge base? I.e. does the system tell the user if a command is not used correctly or constructed correctly?

6. Is it possible to obtain status information about redundant data in the knowledge base? What degree of redundancy will occur in the current product model (0% - 100%)? An example could be the percentage of identical attributes, i.e. attributes which are identically defined twice or more.

7. Can the system deal with an infinite (parametric) solution space, or can it only handle a finite (combinatorial) solution space?

8. Is there a maximum number of classes (objects), attributes or methods that can be included in the software?

9. Which programming/ modelling abilities are needed in order to insert knowledge into the knowledge base?

Explanatory system

To what extent can the explanatory system in the expert system provide information about conflicts, reasons for conflicts etc.? How well does the expert system explain what it is doing?

Default values

Is it possible to use default values in the configurator, so a standard recommendation for all the time-consuming choices can be produced on the basis of just a few typed-in parameters?

Data security

What level of data security does the system have? The security mechanisms must make it impossible for confidential data (such as procedures or sales price calculations) to be revealed outside the company.

Visualization

What degree of visualization does the system deal with? Does it for example offer no visualization, static 2D, dynamic 2D, static 3D or dynamic 3D visualization?

Other criteria include:

1. Is it possible to "drag and drop" objects in the viewer?
2. Which navigation possibilities are there?
3. Can the viewer operate in a browser?
4. Which formats can the system handle – xgl, stl, vrml etc.?
5. How many operations can be performed on an object – translation, rotation around x, y, z, scaling, grouping?
6. How many objects can the system handle in a scene?
7. Is there a snap function, so it is not necessary to be very precise in placing objects?
8. Can the system deal with collision detection?
9. Is it possible to manipulate objects to within millimetres, or do the objects get moved 1 cm at a time?
10. Is it possible to define different "default views"
11. Can the system deal with different light and shadow functions?

Scalability

When the number of users (e.g. salespersons, customers, employees who must maintain the system etc.) is extended, the system must be able to be expanded to handle the increased load, for example by using more servers.

Integration: open systems (API, ODBC, DDE/OLE)

The configurator must be based on a so-called open structure, so the system follows international standards and permits interconnection or combination of software and equipment, independent of supplier.

Portability (mobility)

The system should be evaluated with respect to portability, especially as regards use of different operating systems (Windows NT, UNIX/Linux etc.)

Updating the software

Future updates of the software must be compatible with earlier versions, so unnecessary development work can be avoided.

Software interfaces

User interface (UI)

Are the user interfaces generated automatically or do they have to be programmed? If they have to be programmed, which technology has to be used? Is it for example Java, Visual Basic or .Net?

Multiple user interfaces (personalization)

Can the system handle multiple user interfaces? I.e.can each user be given a particular user interface based on the same knowledge base? For example, can the system deal with different languages (English, German, French etc.), depending on who logs into the system.

Integration with Office products

The development and adaptation work needed for integrating the system into the usual Office products (spreadsheet or word processing systems etc.) should be evaluated. Making/generating documents such as quotes, quality manuals and user manuals are often an important output from a configuration system. It can be necessary to handle different versions (short and long) of e.g. quote documents; but also versions in different languages can be required.

Integration with Back-Office (databases, production equipment, CAD, PDM etc.)

It should be carefully evaluated how a possible integration between the configuration software and the company's Back-Office can be carried out. The evaluation should especially focus on which tools (standard integration modules) can support this task.

Number of users

Documentation for the system's ability to handle the total number of users should be provided by the supplier.

Product complexity (number of objects and rules)

The supplier should describe and provide documentation for how the system handles complex product structures.

Response times

What are the response times for the software? Does some kind of documentation exist for the software's response times in relation to the number of users, number of objects and rules etc.?

Stability

How stable is the system? Does it often crash? I.e. the system should be evaluated for its operational reliability.

Motivation and commitment

Finally, it is important to evaluate the extent to which the system lives up to the company's current and future requirements with respect to user friendliness and quality. User friendliness and quality can help ensure employees' high degree of motivation and commitment to use the configuration system.

Criteria for choosing supplier

In the following, we consider a number of relevant factors for evaluating the software supplier.

Company profile and visions

When choosing a supplier, it is important to find out about the supplier's market profile and the closely related issue of which visions (with respect to technology, application etc.) they have for the product area. Some suppliers have established large development organisations and aim constantly to be at the very front with respect to the exploitation of new technology and corresponding new possibilities for applications.

Market share

It is important to be aware of the IT system's market penetration. A large market share can indicate one or more of the following factors about the supplier:

- large application potential

- well-organised sales and service organisation
- efficient knowledge and training organisation
- healthy and efficient business procedures

Similarly, a low market share can indicate that the supplier has weaknesses in one or more of these areas. When evaluating market share, it is important to distinguish between the units used to measure the market share – for example, the number of implementations, number of users/sold licenses, technology level, R&D budget, turnover etc.

Product development (R&D)

It is relevant to measure/evaluate the supplier's R&D organisation (development staff and development budgets, to the extent these are available). An extensive R&D organisation and corresponding budgets help to position the supplier in the market. One of the reasons a company chooses a standard system rather than developing the product itself is precisely the extensive product development carried out on standard systems. The company can expect continued development of the product, based on experience from a large number of companies and users.

Quality

When choosing a supplier, the quality of the system should be clearly specified. Similarly, it must be clear which areas can be critical but necessary to accept. The supplier should supply documentation demonstrating that the system can live up to the specified quality standards for operational security, data reliability, performance etc.

Service/support

A company purchasing standard software for configuration purposes wants professional and efficient IT support; therefore, it is important to evaluate the supplier's service organisation with respect to experience, resources and use of technology (e.g. whether errors are to be reported by telephone, e-mail or via the Internet). If the supplier collaborates with partners to solve service/support tasks, then it is also necessary to evaluate the actual partner/sub-contractor on the basis of the criteria listed in this section.

Training

It is also of great significance that the supplier has the resources and the competence to carry out training within the company. Otherwise, acceptable alternatives should be investigated – for example the use of partners.

Resources

As can be seen from several of the points mentioned above, it is important that the necessary resources can be procured from the supplier for carrying out the implementation project. If these are not available from the supplier, then it is necessary to use the supplier's partners, if any.

Trade knowledge and trade-specific solutions

Certain suppliers are specialized within a particular industry or branch, and therefore have specialized solutions adapted to suit this industry.

References (number of installations/licenses/users, user groups)

A commonly used and reliable indicator for a supplier's position in the market is the number and size of systems it has delivered. A list of references can also show the number of installations with corresponding users (licenses) and user groups. It can be of considerable significance, if it is possible for the company to visit some of the reference companies provided, in order to evaluate the use of the technology in practice.

Geographical position

When purchasing a standard system, it is essential that the necessary assistance and support for implementation, maintenance and servicing (hot-line activities) are available. It is to be expected that such services can be delivered from within the country concerned, or at least from neighbouring countries. If the company has international branches, the requirement can be extended to cover several countries or regions.

Price (product, implementation, service, maintenance and training)

The suppliers must have an unambiguous price structure specifying separately prices for software, implementation services, support for maintenance, and services for facilitating the necessary training. It must be clear whether a delivery has a fixed price or whether the price is based on the time used for implementing the project.

License conditions

The price may for example be determined on the basis of the number of licenses purchased; therefore, it is important to gain insight into and understand the supplier's licensing rules (i.e. does the price depend on the number of users, number of servers, number of objects, number of visitors to a webpage, number of development environments etc.)

Examples of programming expert systems

In this section, we present two examples of programming of expert systems: "programming in non-object-oriented expert systems" and "programming in object-oriented expert systems".

Programming in non-object-oriented expert systems

If expert systems cannot handle an object-oriented knowledge representation in the knowledge base, they often have a "flat" structure such as the expert system shown in figure 8.6. By a "flat" structure we mean that classes, attributes and methods (rules) do not have any hierarchical structure or do not follow the basic object-oriented principles. An expert system which does not support object-oriented principles does not handle e.g. class-object relationships, inheritance, or encapsulation. In this presentation, this is called a "non-object-oriented expert system".

Window 1: Contains attributes and their possible variations, resources and calculations. They can be grouped via folders.

Window 2: Project control (models)

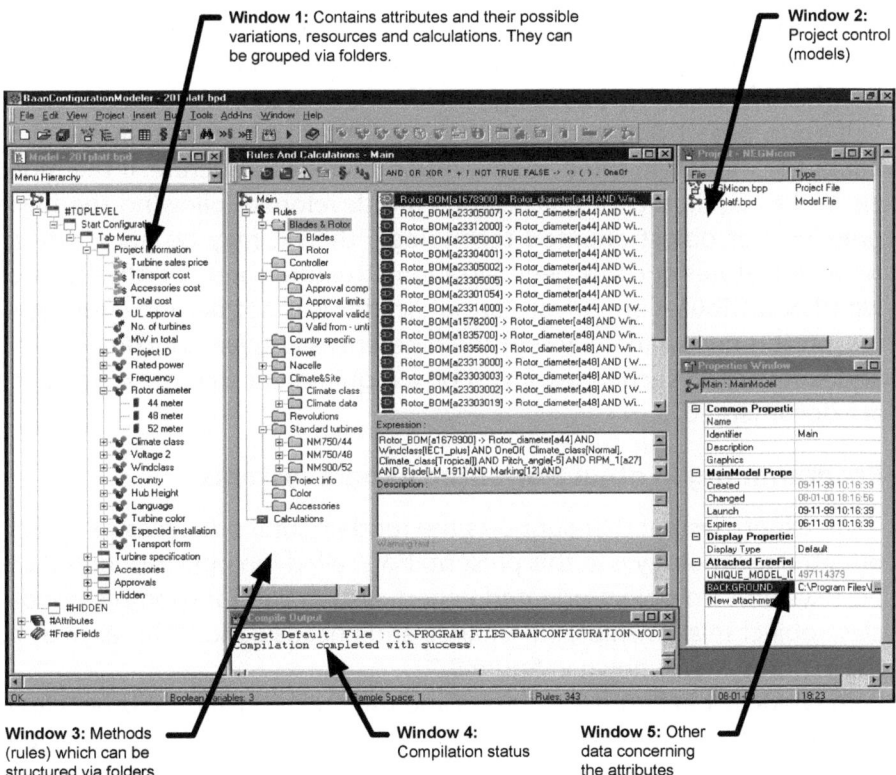

Window 3: Methods (rules) which can be structured via folders

Window 4: Compilation status

Window 5: Other data concerning the attributes

Figure 8.6. *Constraint-based expert system with a "flat" structure.* © Copyright 2006. Infor Global Solutions. All Rights Reserved.

In figure 8.6 under Window 1, the attributes (the data types are denoted "Single", "OneOf", "AtMostOne" etc. by the software supplier) and their possible variations (denoted elements) can be grouped in a series of folders (called menus). A number of resources and calculations can also be created. The resources make it possible for example to associate a price or an assembly time with the attributes. The calculations make it possible to add, multiply, divide etc. In Window 3, the methods (rules) are set up on the basis of the attributes, resources and calculations set up previously. The methods can be grouped using folders to make it easier to get an overall view, even if the actual structure for the methods is "flat".

In a non-object-oriented expert system, it is therefore not possible to create classes which are encapsulated or to use generalization structures. Therefore, the structure in these systems (e.g. for attributes and methods) is determined solely on the basis of which folder structure is chosen in creating the knowledge base. The structure in the folders must as far as possible reflect the structure in the models that are chosen as documentation for the product (the product variant master, class diagrams, CRC cards etc.) The folders can be structured according to cause-effect relationships, function types, part types, life phase systems, properties etc.

If a non-object-oriented expert system is used, it is often convenient to adapt the models in the documentation to a "flat" structure. If the models in the requirements specification are developed following the basic principles of object-oriented technology, then it may be necessary to remove features such as generalization structures and specifications of whether attributes are private or public. If this is not desirable for some reason, it is necessary to accept the differences between documentation and software. In such cases, it can be a requirement that the differences are documented.

Programming in object-oriented expert systems

An expert system that supports the fundamental principles of object-oriented technology is in this presentation called "an object-oriented expert system". An object-oriented expert system can for example handle class-object relationships, inheritance and encapsulation. This applies to the expert system shown in figure 8.7.

In Window 1 in figure 8.7, classes and their relationships to other classes (such as an aggregation structure) are defined. In Windows 2 and 3, the classes' attributes (including variation possibilities) and the classes' methods (rules) are defined. The relationship between Windows 1-3 and the notation for a class is illustrated in figure 8.8.

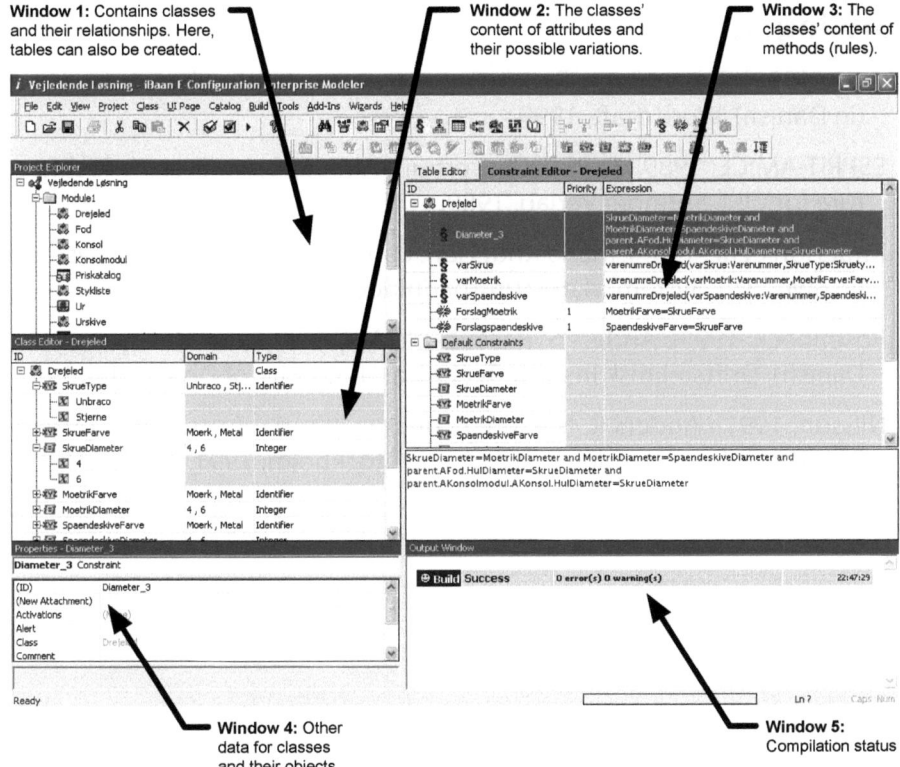

Window 1: Contains classes and their relationships. Here, tables can also be created.

Window 2: The classes' content of attributes and their possible variations.

Window 3: The classes' content of methods (rules).

Window 4: Other data for classes and their objects.

Window 5: Compilation status

Figure 8.7. *Constraint-based expert system with object- oriented structure.*

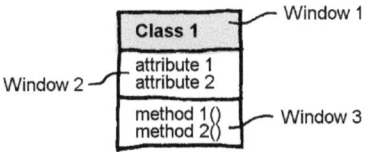

Figure 8.8. *Relationship between the expert system's windows and the class notation*

It is often best to use an object-oriented structure in the knowledge base, since this makes maintenance easier, but this is unfortunately not particularly widespread in the standard expert systems existing today. Standard systems often only support a few or none of the basic object-oriented principles.

Bibliography

[Buch, 1999]: Tobias Buch; Requirement Specification - sales configurator (in Danish), Work draft, 1999.

[ESPRIT-AMICE, 1989]: ESPRIT Consortium AMICE; Open System Architecture for CIM, Springer Verlag, 1989.

[Knutsson et al., 1997]: Maria Knutsson, Carina Franzén; Standard systems (in Swedish), Systemarkitektururbildningen, 1997.

[Rosenbeck, 1995]: Anders S. Rosenbeck; Acquiring standard systems (in Danish), Driftsteknisk Institut, Technical University of Denmark, 1995.

[Nilsson, 1991]: Anders R. Nilsson; Acquiring standard systems (in Swedish), Thesis from Handelshögskolen in Stockholm, 1991.

9

Product Configuration at F.L. Smidth

This chapter describes how F.L. Smidth A/S has developed a configuration system for general dimensioning of cement factories and working out budgetary offers. After a short introduction to F.L. Smidth A/S, the company's products, and their project for building up a configuration system, the chapter follows the procedure for developing configuration systems as described in chapter 3.

The company F.L. Smidth A/S is part of the F.L. Smidth Group, which in turn is part of the FLS Industries concern. In 2003, F.L. Smidth A/S, which is the largest company in FLS Industries, had a turnover of 687 million euro and 2109 employees.

F.L. Smidth A/S supplies complete factories for cement production. They also supply parts of factories and renovate existing cement factories. Figure 9.1 shows a cement factory delivered to Sinai in Egypt.

A cement factory consists of a series of departments, such as for storing clay and limestone, grinding raw materials and coal, an oven for firing clinker, a department for grinding of clinker, and for storing and packing finished cement. Cement factories are designed according to the characteristics of the raw materials, capacity requirements, emission requirements etc. A complete cement factory typically costs about 140 million euro, and the lead time for designing and building a complete factory is 2-3 years.

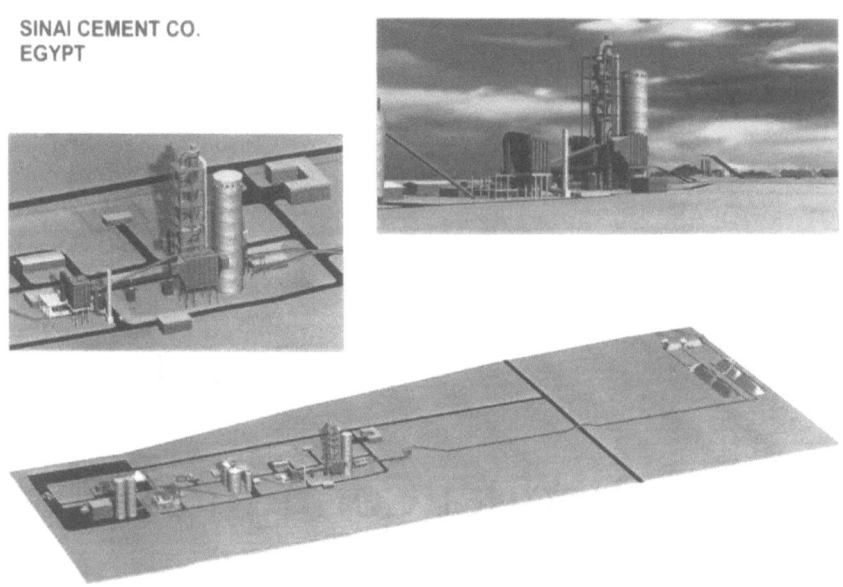

Figure 9.1. *A cement factory.*

F.L. Smidth currently has a market share of more than 30% (measured in oven capacity). F.L. Smidth's biggest competitor is Polysius, which has a market share of about 15-20%. The remainder of the market is divided among a number of smaller competitors.

The customers are very diverse. They can be private persons with substantial capital who have not previously worked with cement production, states or public bodies, or current producers of cement. The most important customer requirements for cement factories can be summarized as:

- Price and financing conditions.

- Delivery time. Short delivery time reduces the risk connected with whether the cement produced will be able to be sold. Quick delivery also means that the cement factory can be put into operation early, and therefore more rapidly contribute to repaying the invested capital. Loss of production time due to long delivery time for the cement factory means lost sales for F.L. Smidth's customers.

- Operating costs. The running costs associated, for example, with the workforce, energy, transport, maintenance etc. are critical for the cement factory's overall rentability.

- Energy consumption and environmental load (emissions). In recent years, there has been increasing focus on minimizing energy consumption and emissions from cement factories.

The market for cement factories has been declining in recent years. F.L. Smidth experiences an increasing pressure from the market to be able to produce a binding offer. At the same time, the company uses a large - and increasing - portion of its resources to produce offers. This is the background for F.L. Smidth's decision to initiate a project to develop a configuration system for working out budgetary offers.

Phase 1: Analysis of specification processes

Identification and characterization of the most important specification processes

The first step in the project to develop a configuration system is to identify the most important specification processes within F.L. Smidth. It was agreed from the start that the production of offers as part of the sales process was of critical significance for the company. The reason given for this was that it is in the process of making sales and offers that the company meets the customers and the projects are created – without sales, no company. It is also in the sales process that all important parameters for the cement factory are defined. As described in chapter 4, the most important decisions about the cement factory are made early in the project. The critical performance parameters for the cement factory, and all the most important cost parameters, are already decided at an early stage in the offer process. Therefore, the decision was made to focus on the early phases of the offer process.

To be able to estimate the potential offered by the use of product configuration for supporting the task of making offers, it was necessary to make an analysis of the current offer process, a summary of the most important aims for the offer process – compared with the current performance, and a vision for how a configuration system could contribute to making offers in the future. Considerable emphasis was also placed on setting the limits for the configuration system, so that it was certain that the project could be implemented.

In connection with the analysis of the current offer process, it turned out to be rather difficult to describe the process of working out budgetary offers. The process varies from one offer to another, depending on the nature of the offer or project, and on the persons involved in the work.

The big differences in the process of working out a budgetary offer also mean that the offers produced can be very different and of varying quality, depending on who has been responsible for making them.

Normally, making offers involves staff from several departments, and considerable resources are used to coordinate and check the information flowing between the different departments.

Offers at F.L. Smidth are divided up into two main groups: budgetary offers and detailed offers. Normally, the customer first receives one or more budgetary offers. Then, when the customer has reached a decision about the general requirements for the cement factory, one or more detailed offers are worked out. A frequent problem is that the customer changes the general requirements for the cement factory quite late in the detailed-offer phase. This means that the customer must often be provided with several of the time-consuming detailed offers before the order is placed.

A budgetary offer contains, amongst other things, a general description of the cement factory, with a specification of operational factors such as capacity or emissions, a process diagram, a list of the cement factory's principal machines, a price calculation, a list of auxiliary equipment and a timetable. A budgetary offer is typically 100-200 A4 pages long and can take 1-4 weeks to make, with a resource consumption of about 5 man-weeks.

A detailed offer is considerably more detailed than a budgetary offer and includes a complete and detailed description of all the cement factory's departments with specifications of all the buildings, machines, controls etc. In addition, it contains a detailed plan for the cement factory's construction on site, and the factors related to the factory's initial commissioning and operation. A detailed offer typically fills 10-20 ring binders and can take 3-6 months to produce with resource consumption corresponding to 1-3 man-years.

Targets and requirements for the offer process compared with current performance

Figure 9.2 summarizes some of the most important targets for the process of working out budgetary offers. The targets are compared with current performance.

As can be seen from this gap analysis, the use of a configuration system for producing budgetary offers will cause considerable reductions in lead time and resource consumption. This means that it would be possible to respond to all enquiries with a budgetary offer. Without a configuration system, the staff at F.L. Smidth are forced to decide which customer enquiries they should give highest priority, since they do not have enough resources to answer all enquiries with a budgetary offer. Investigations show that it is exceptionally difficult to judge whether an enquiry will result in an order. Thus, some of the past enquiries that never received an offer in response have become customers of one of F.L. Smidth's competitors, who have run off with the order. It is therefore of vital importance for F.L. Smidth that all enquiries receive a responce in the form of a budgetary offer.

	Target	Current performance	Gap
Lead time	1-2 days	5-25 days	To be reduced by about 50-90%
Resource consumption	1-2 man-days per budgetary offer	About 25 man-days per budgetary offer	To be reduced by about 90%
% enquiries which are replied to with an offer	100%	50%	50%
Quality of budgetary offers	Uniform budgetary offers	Large variation	More uniform offers

Figure 9.2. *Gap analysis of F.L. Smidth's process for working out budgetary offers.*

The budgetary offers become more uniform and of better quality in the respect that all relevant factors are taken into account and the difference is reduced between the costs calculated for the budgetary offer and the actual costs. The offers generated by the configuration system are targeted at business people, i.e. the focus is on the general principles, the cement factory's performance and the technical factors that affect the price and/or the cement factory's performance.

It also becomes possible to optimize the cement factory, for example with respect to the use of previously designed and produced parts in the factory and to the use of parts produced within the FLS concern.

One of the most important gains from a configuration system will be to guide F.L. Smidth´s customers in the direction of choosing F.L. Smidth´s standard solutions, rather than solutions that are tailor-made to match customer needs which – when seen in relation to the overall project – can often be quite unimportant for the customer. But which lead to a markedly increased work load, with resulting delays and uncertainty for the entire project. F.L. Smidth's experience is that if F.L. Smidth can offer a standard solution with a fixed delivery time and price rather than a customer-specific solution, which would be more expensive and take longer, then the customers will in the great majority of cases choose a standard solution.

Finally, F.L. Smidth's staff sees the configuration project as a way of formalizing and sharing knowledge about the company's products. The construction of the product variant master, class diagram and CRC cards provides a unique possibility for creating an overall picture of the total scope of the cement factories F.L. Smidth wants to offer the market.

Scenarios for use of configuration at F.L. Smidth

The future process for producing offers at F.L. Smidth is connected with a vision of developing a configuration system to assist in making budgetary offers. By using such a configuration system, it should be possible to make a budgetary offer in 1-2 days, which is also detailed and precise enough to be able to replace some of the detailed offers produced today.

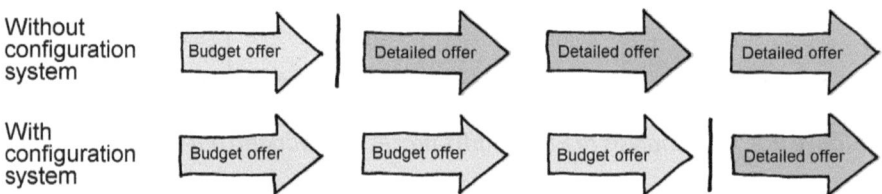

Figure 9.3. *The configuration system leads to fewer detailed offers.*

In other words, the overall vision for future offer production at F.L. Smidth is of a process in which, by using a configuration system, it is possible to make budgetary offers quickly and efficiently, so that a customer can receive, for example, 5-10 different budgetary offers and then, when the cement factory's overall dimensions have been determined, one detailed offer. In this way, F.L. Smidth expects to save resources on working out budgetary offers, at the same time as fewer detailed offers have to be worked out than before.

Figure 9.4 illustrates how the configuration system is used in the process of producing budgetary offers. The configuration system can be used by a salesperson, who determines all the necessary inputs for the configuration system together with the customer and then prints out a budgetary offer. Another procedure could be that a group of F.L. Smidth's sales staff and mechanical specialists, on the basis of an enquiry from a customer, determine together the necessary input for the configuration system, perform a configuration, and print out a budgetary offer. This is then checked for possible mistakes before being sent to the customer.

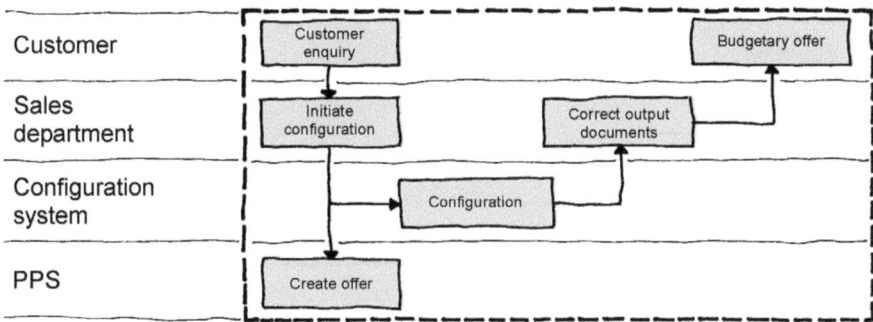

Figure 9.4. *Working out budgetary offers by means of a configuration system.*

Figure 9.5 shows the overall process with budgetary offers and detailed offers after introduction of a configuration system. Detailed offers are worked out on the basis of the budgetary offer(s) previously produced for the customer. Since the budgetary offers using a configuration system become more uniform and have higher quality, this will also contribute to increasing the quality of the detailed offers. It is also expected that the configuration system will be able to contribute to working out some of the detailed offers. In the long term, the idea is to develop a set of config-uration systems for the most important main machines that will be able to support both detailed offers and order processing.

Figure 9.6 shows how the use of modules and the configuration system contribute to the overall process up to signing the contract. In the figure, a "feasibility study" phase has been included. This consists in working out a budgetary offer from just a few items of customer input, based on FLS standards and default values.

Modules and concepts are described in the form of a set of drawings with specifications of the most important functional properties, such as capacity or emissions, together with a list of the equipment included in the individual modules. In addition, the modules are represented in the configuration system, together with rules for calculating operating times, capacity etc., and rules for selecting and dimensioning the individual modules. The configuration system also contains a series of typical layout drawings and flowsheets with the associated rules for adapting them. The

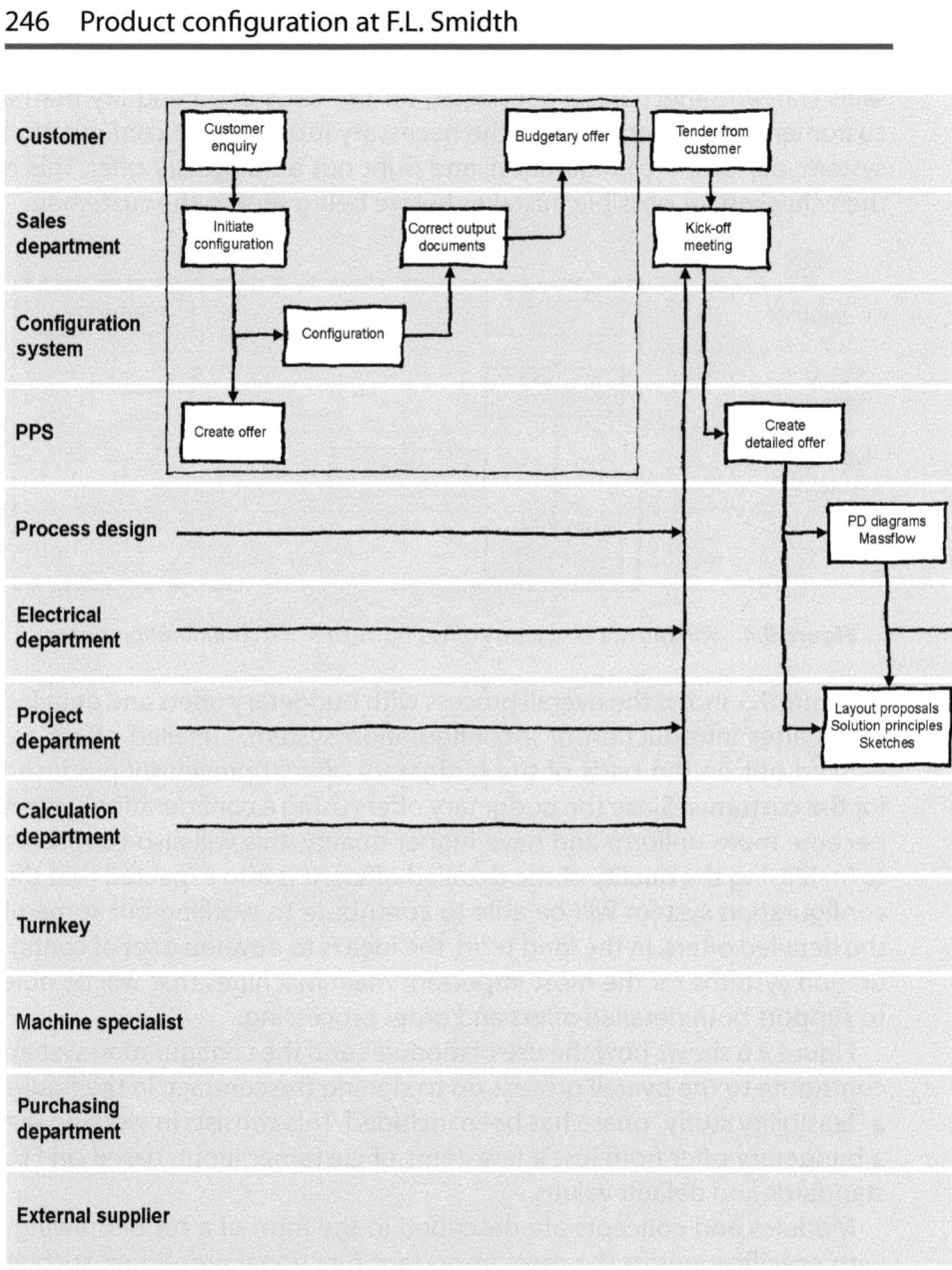

Figure 9.5. *Budgetary offers and detailed offers.*

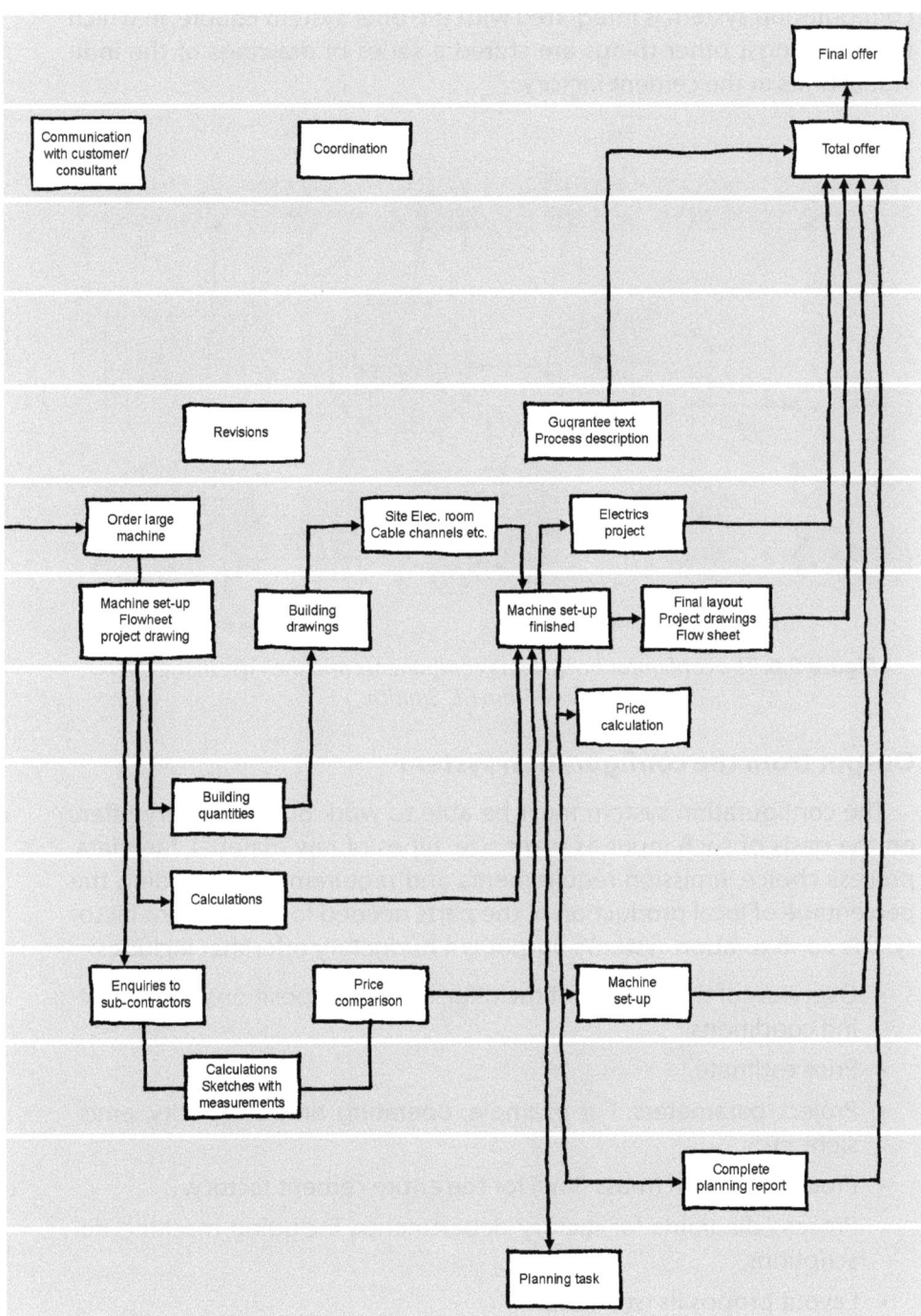

configuration system is integrated with the offer system Equote, in which there amongst other things are stored a series of drawings of the individual units in the cement factory.

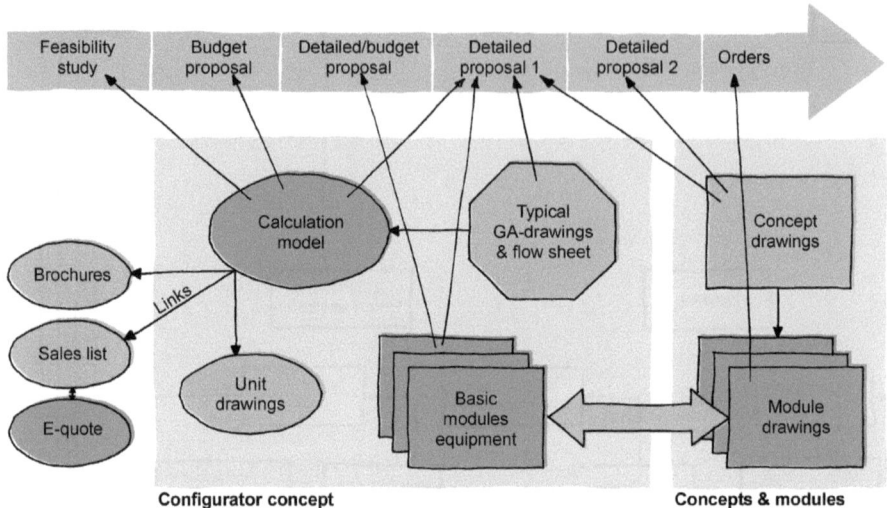

Figure 9.6. *The configuration system's components and their use in the offer process. From F.L. Smidth.*

Output from the configuration system

The configuration system must be able to work out budgetary offers on the basis of such input as plant size, types of raw material, fuel data, process choice, emission requirements and requirements regarding the percentage of local production of the parts needed for the cement factory. The configuration system's output is a budgetary offer that includes:

- Overview of the content of the offer, its pre-suppositions and financing conditions.
- Price estimate.
- Project parameters. For example, operating times, capacity, emissions etc.
- Process diagram (mass flow) for the entire cement factory.
- Process diagrams for factory departments, including machine descriptions.
- Layout proposals (standard).
- Description of main machines, including sketch of principles.

- General description of electricity supply and controls.
- Timetable.

The budgetary offer is created as a Word document. A macro is programmed in MS Word that accepts input (text and drawings) related to the individual elements in the configuration selected from the configuration system. A budgetary offer is typically about 100-200 pages long.

The future process for making offers

Three different scenarios for using product configuration at F.L. Smidth are described below.

The first scenario has a configuration system that can perform high-level dimensioning of a complete cement factory and work out a process diagram and a letter with an offer. The configuration system must deal with cement factories situated up to 200 m above sea level. In addition, only those parts of the cement factory for which the most demand is expected are included. The configuration system includes only the actual cement factory's design and function. In other words, life-cycle properties such as the manufacture of the cement factory, or its operation and maintenance are not included in the configuration system.

The second scenario is more extensive and includes more variants, such as silos, filter systems, grinding mills and the principle of mixed stores, as well as cement factories placed more than 200 m above sea level. In addition, the configuration system must include the building process by calculating the cubic metres of concrete needed for the various departments. This is for example of great importance when configuring storage, since the actual building represents a larger part than the machine part. It must also be possible to include prices and weights for parts of the cement factory that the customer wants to obtain from suppliers other than F.L. Smidth. With respect to life-cycle properties, the configuration system in the second scenario must include a calculation of production costs in the country where the individual parts of the cement factory are to be produced.

The first scenario involves configuration of a complete cement factory. In the second scenario, it must also be possible to configure parts of the cement factory, such as coal mills or clinker grinding stations. For those parts that must be able to be configured separately, the degree of detail has to be so high that it is possible to calculate an exact price for the individual parts.

The third scenario involves configuration of both the complete cement factory and all the individual departments. In addition, special configuration systems especially adapted to suit the needs of the five largest customers are produced. In the third scenario, it must also be possible to produce a list of the machine layout to be used as the basis for making detailed offers. With respect to life-cycle properties, the possibility for delivering spare parts, and training and/or consultancy in connection with the commissioning and operation of the cement factory must be included.

A model of the complete cement factory is produced by combining the selected departments and machines in a complete 3D model, so the customer is able to experience a virtual walk-through of the entire cement factory. This will markedly improve the customer's ability to obtain an overall view of the cement factory, and this will make clear the consequences of the different choices of machines made in the configuration system.

Figure 9.7 illustrates the three scenarios for using configuration at F.L. Smidth. The three scenarios describe the development of the configuration system in time. Scenario 1 is the configuration system implemented in 2000; scenario 2 is the configuration system implemented in 2005; while the third scenario describes the target set for development of the configuration system in the period up to 2010.

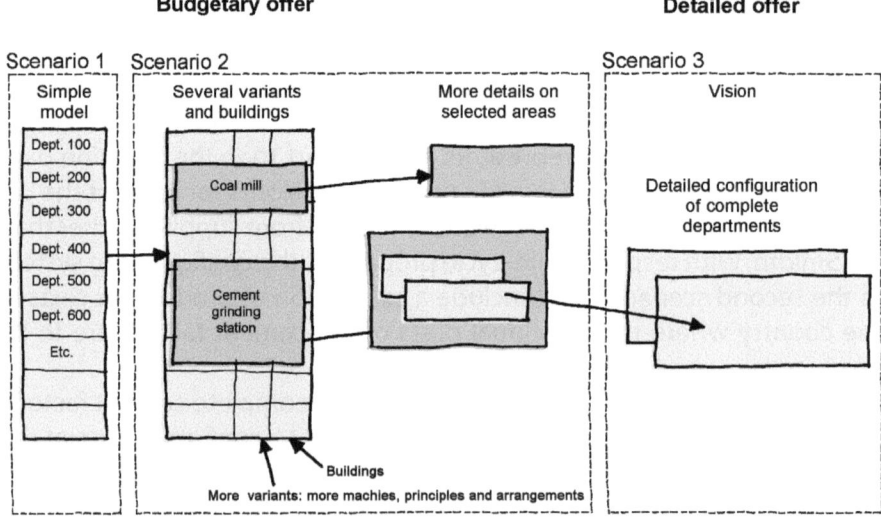

Figure 9.7. *Three scenarios for the use of product configuration at F.L. Smidth.*

Definition of the configuration system

The configuration system to be built is defined, and its scope determined by using the framework system for modelling product families described in Chapters 2 and 4.

In the configuration system, the main focus is on the machines, since an important share of the cement factory's price is dictated by the machines. The remaining technical disciplines are only considered to the extent that they affect the price considerably. The building part, which makes up a considerable share of the cost of the cement factory (about 40%), is included in scenario two of the configuration system, based on a calculation of the expected use of concrete and steel reinforcement.

In order to dimension a cement factory, we start with a general list of requirements for the cement factory expressed in terms of a scope list, which includes a description of the capacity requirements, emission limits, heating plant, and raw material characteristics. This is followed by a general description of the process followed in the cement factory (mass flow). On the basis of the scope-list and the high-level process description, solutions in principle for the individual sections of the factory are chosen. This could for example involve a choice of the principle for storing raw materials, which could be between longitudinal storage, side scraper storage or circular storage. The individual solutions in principle are described by a process diagram and the main machines involved in the solution. In connection with the choice of a solution in principle, a layout diagram is made for the relevant section of the factory, which shows where the machines are physically placed in relation to each other.

When the solution principle has been chosen, then the main machines involved in this solution, and their dimensions, are selected. The main machines are drawn up in so-called type rows, which are lists of machines ranked in order of capacity. The machines included in a type row have not all been used previously at F.L. Smidth; therefore, when choosing the machines in a type row, it must also be considered whether a set of drawings and some experience in using the machine concerned are available.

The scope list and the high-level mass flow diagram describe the functional requirements for the cement factory, while the solutions in principle can be considered as function-bearing elements that describe different ways for achieving a given function. The description of the machines involved, ordered in type rows, can be considered as a description of the cement factory's component parts, i.e. a part-structure.

Figure 9.8 illustrates, using the framework system for structuring product knowledge presented in chapter 2, the overall content and structure

	Property models (Derived properties)		Product structre model		Models of the product's meeting with life cycle systems		
	In- and external properties	Functional properties	Solution principles	Part model	Production model	Installation and commissioning	After sales service
Generic level	Describe cement factory's other properties	Describe cement factory's function	Describe solution principles	Describe cement factory's main machines	General characteristics of production in different countries	Describe FLS services in connection with installation and commissioning of cement factory	Describe FLS services related to supply of spare parts
	Rules for calculating price and weight, and for specifying whether individual parts of the cement factory have previously been designed and produced	Rules for calculating capacity, energy consumption, emissions etc., and relationships to solution principles	Rules for selection of solution principles	Rules for selection of machine size	Rules for adjustment of production price according to production country	Rules for selection of FLS services for installation and commissioning of cement factories	Rules for selection of spare parts
Instance level	Price tables, weight tables, and table of which main machines have previously been designed and produced	Tables with capacity, energy consumption and emissions for solution principles and main machines	Description of solution principles	Type rows	Tables of characteristics of price, quality, delivery conditions etc. for production in different countries	Tables of FLS services for installation and commissioning of cement factories	Tables of prices and specifications for spare parts
	Scenario 1				Scenario 2		Scenario 3

Figure 9.8. Framework with description of the overall contents and structure of the configuration system.

of a configuration system to use in preparing budgetary offers. In the framework system, the types of knowledge included in the three scenarios are also shown.

Apart from parts, solutions in principle (function-bearing units) and functions, the framework also includes the properties of price and weight. Under properties, information about whether the part of the cement factory concerned has previously been designed in detail and delivered is also included, as this has great influence on both costs (engineer hours at F.L. Smidth) and delivery time.

Capacity, energy consumption and emissions are modelled under functions, as they contribute to describing the cement factory's overall function. The configuration system in scenario 1 focuses on the actual cement factory's construction and function. In connection with scenarios 2 and 3, knowledge which describes the cement factory's production, installation and commissioning, and its operation and maintenance, is included. In other words, in scenario 2, calculation of the price in different countries is included, while scenario 3 includes information about commissioning the cement factory and maintenance and operation, together with sale of spare parts and replacement of parts.

The production price has been included in scenario 2, because it is also necessary, in the context of working out budgetary offers, to estimate the costs of producing the cement factory. The manufacturing costs are included in the estimated prices for the various parts of the cement factory; however, production costs are not constant throughout the world. In order to find a sufficiently accurate price, it is therefore necessary to consider where in the world the various parts of the cement factory are to be produced.

This is done by specifying how large a percentage of the total manufacture is to be carried out in a particular country, and then correct the price in relation to cost levels in the country in question.

In connection with construction of the cement factory on site, installation, and commissioning, it is necessary to decide how large a part of these activities the customer wishes to perform and which activities are to be performed by F.L. Smidth. F.L. Smidth offers such services as construction, commissioning, and consultancy or training the customer's operations engineers. The extent of these services must be determined in the configuration system and be included in the calculation of the budget price. This is included in scenario 3.

In connection with the operation and maintenance of the cement factory, F.L. Smidth offers a series of services such as service contracts, where

F.L. Smidth takes on the task of maintaining all or part of the cement factory, training the staff, or delivering spare parts.

The various services are included in the configuration system in scenario 3. The various services are listed, and prices for the individual services are included together with rules for how they can be combined.

Plan of action for carrying out scenario 1

After the first prototype was developed, a plan of action was produced for construction and implementation of a configuration system, corresponding to scenario 1. The task of constructing and implementing the system involved selection of configuration software, construction and documentation of a product model, production of prototypes for selected areas to ensure that the system could deal with the task, and ongoing tests with selected users. The main points in the project's workplan were as follows:

- Description of the future offer process when using the configuration system. Final delimitation and definition of the configuration system. Definition of the configuration system's user interface.
- Construction and documentation of the product model with the help of a product variant master and CRC cards.
- Selection of configuration software. Selection of relevant systems. Test of systems. Selection of system and finalizing the contract with supplier.
- Programming and test of critical parts of the configuration system.
- Development of system in Visio for drawing the process diagram, together with macros in MS-Word for production of written offers.
- Integration with F.L. Smidth´s price database etc.
- Programming of the configuration system.
- Test and correction of the configuration system.
- Training of the configuration system's users.
- Use and test of the configuration system by the configuration system's users.

A total of about 800,000 euro were used in the phases up to and including the implementation of the configuration system in the year 2000. Running costs make up about 1-2 man-years per year.

Organisation of work in scenario 1

The execution of scenario 1 is organised as shown in figure 9.9. The project's sponsor is F.L. Smidth's technical director. The project is managed by a project leader and a facilitator from the Technical University of Denmark (DTU). The task of change manager is the responsibility of the project leader, together with the technical director. The project leader and another employee are the model managers, while the project leader also has responsibility for developing the future offer process using a configuration system. The mechanical specialists at F.L. Smidth, together with the project leader, fulfil the roles of domain experts. The configuration system is programmed by a staff member who has both an IT and engineering background.

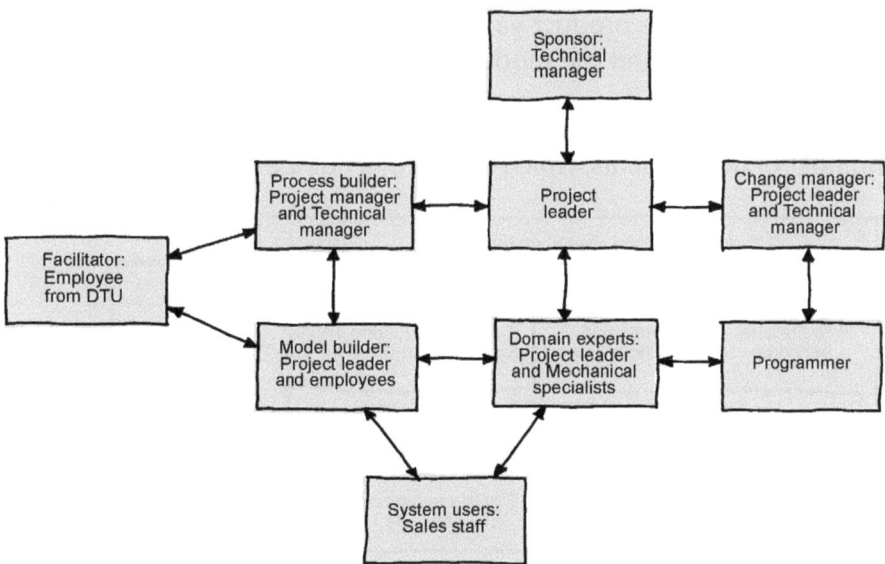

Figure 9.9. Organisation of the work involved in executing scenario 1.

The configuration system (scenario 1) was put into operation in 2000. In 2003, a development project was started to realize scenario 2. The first step in this development project was a preliminary project, in which a detailed description of the content of scenario 2 was developed, with a specification of the future offer process and the scope and definition of the new configuration system. In addition, one part of the product model was produced with a product variant master and CRC cards, together with a prototype of the future configuration system using the configuration system eConfiguration Enterprise from Infor.

After this preliminary project, the full project of realizing scenario 2 was implemented by a staff member who acted as project leader, change manager and model manager, together with a model builder and a programmer, in collaboration with mechanical specialists (domain experts) and sales staff (users of the system).

The product model and the configuration system shown in the next section are from scenario 2.

Phase 2: Product analysis

The product analysis was carried out using the scope, which was described above in connection with the use of the framework, as its starting point. The product analysis was performed by drawing the various parts of the cement factory in a product variant master. Before describing the work of making the product variant master, we describe how a cement factory is built up and the module structure for cement factories developed at F.L. Smidth.

A cement factory consists of process sections, which in turn consist of a number of departments, which consist of a number of machines.

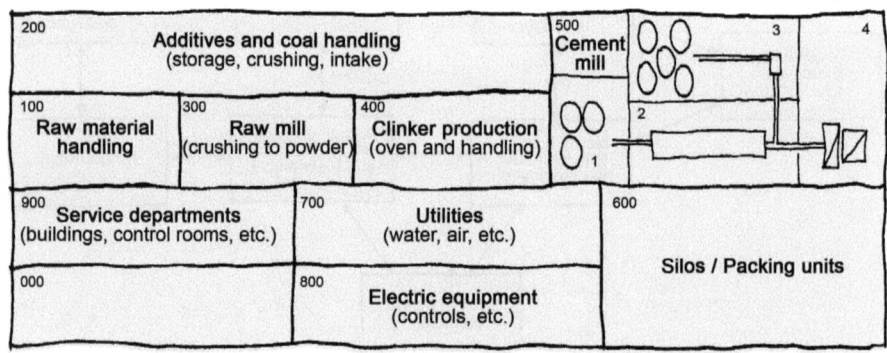

Figure 9.10. *Process sections and departments in a cement factory.*

Figure 9.10 shows how a cement factory can be broken down into process sections, departments and machines. Process sections are defined by the numbers 100-900. The individual process sections are defined by their function, capacity etc. There is also a well-defined interface between the individual process sections. For example, the raw-mill grinds raw materials to powder before they are burnt to clinker. The interface between the raw-mill and the oven is defined by the pipe system for hot and cold air to and from the raw-mill, and by the number of cyclonic transport systems, which can be 2 or 4. An example of a process section is section 500, which deals with the grinding of cement clinkers. Process section 500 consists of the departments:

1. Feed silos
2. Cement mill
3. Separator
4. Drive mechanism for the cement mill (motor, gears etc.)

Different solution principles are available for the individual departments. The individual solutions in principle are described by their function, capacity etc., and by the machines involved.

In addition to being divided up into process sections and departments, the cement factory is divided up according to different technical areas/disciplines. Figure 9.11 shows the cement factory's division into process sections and departments, and according to technical areas.

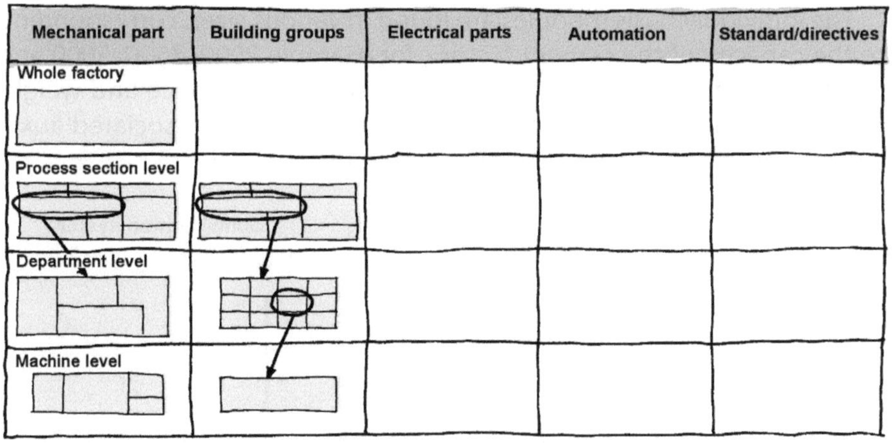

Mechanical part	Building groups	Electrical parts	Automation	Standard/directives
Whole factory				
Process section level				
Department level				
Machine level				

Figure 9.11. *The cement factory can be described in terms of process sections, departments, machines and technical areas.*

The configuration system relies on the use of basis modules. A basis module is a collection of machines and equipment which perform a function in the cement factory, for example storing coal or grinding cement clinkers. The idea of using basis modules is to specify the main machines, as these define the cement factory's capacity etc. and determine 80% of the price of the cement factory.

The basis modules are specified in detail with designated machine lists and auxiliary equipment. In the configuration system, a basis module is defined by the module's function, capacity, price and other factors such as emissions, operating times, energy consumption etc. The configuration system also contains rules for how the individual basis modules can be combined.

Thus, the configuration system only deals with the main machines defined as basis modules. Equipment that connects the basis modules together, such as conveyor belts, blowers etc. are not included in the configuration system, as these parts are not critical for the cement factory's price or capacity.

An example of a basis module is limestone storage, department 131, LHO (Longitude Homogenization Storage), which consists of an incoming belt, the storage unit itself, and an outgoing belt. For this type of storage, a basis module is defined for the incoming belt, a basis module for the outgoing belt, a basis module for a long LHO storage unit, and one for a short LHO storage unit. Basis modules are also defined in the same way for other storage types, such as sideways scraped stores and circular stores.

The individual basis modules are found in various sizes, corresponding to the capacity of the cement factory, for example 2000, 3500, 5000 and 7500 tons per day. On the basis of the basis modules, price and weight curves are worked out for each main machine and the associated auxiliary equipment, if any, as shown in figure 9.12.

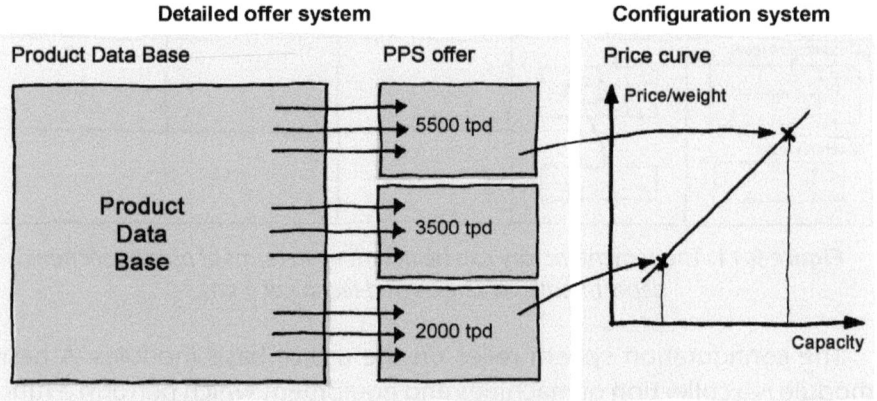

Figure 9.12. *Price and weight curve for main machines.*

A price curve is based on previously produced main machines. The price and weight curves are made by inserting the capacity, price and weight for previously produced machines of the type concerned (typically 3-5 machines). A curve is then drawn through the points, so it is possible to read off the price and weight for machines that have not previously been produced.

The main machines in the basis modules are stored as lists of machines in the product database, which contains information about all the parts included in previously produced cement factories, with references to drawing number, lists of parts, assembly specifications etc. The list of machines is then priced in the offer system (PPS), which is continually being updated with respect to materials prices, exchange rates etc.

The use of basis modules and price curves makes it possible, by using a configuration system, to combine and dimension the basis modules, and thus to put together a suitable cement factory in relation to the available raw materials, the desired capacity, the operation time, emissions etc.

Product variant master

To proceed with the detailed description of the individual parts of the configuration system, a product variant master is used, as described in Chapters 3 and 5.

The product variant master is delimited and structured according to the overall structure given in the framework shown in figure 9.8. The structure of the product variant master is as follows:

- The cement factory's overall functional characteristics and other properties. For example, its capacity, fuel, altitude, and characteristics of the raw materials, and the finished cement.
- Materials flow. Characteristics of the materials flow through the individual process sections.
- Mass flow. A description of the high-level process with a specification of the capacity and other operating conditions.
- Departments. A description of each department with a specification of capacity and other operating conditions, and the relation to mass flow, layouts and choice of machine type.
- Life cycle properties including running costs together with spare parts and replacement of worn parts.

Figure 9.13 shows a small section of the product variant master. This section is from the part of the product variant master describing departments. The section shows the treatment of raw materials and includes a characterization of the raw materials used (such as limestone and shale), conveyor belts and crushers.

In connection with the construction of the product variant master, a series of discussions were held in order to fix the scope of the project and decide on the structure and content of the configuration system. The discussions were partly focused on clarifying the customers' most important requirements for the configuration system, and partly on how the scope and structure of its knowledge about cement factories should be delimited and structured.

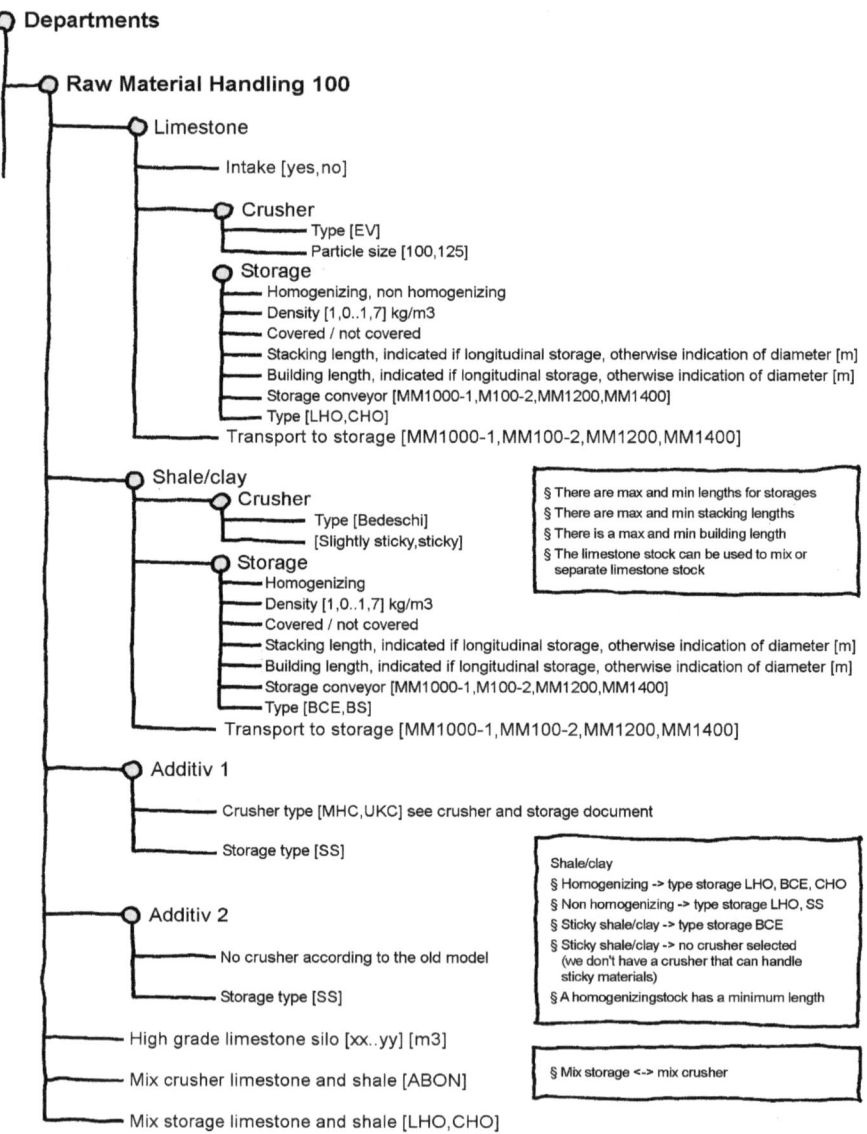

Figure 9.13. *A section of the product variant master.*

One of the results of these discussions was the decision to employ a "material-flow point of view" in constructing and structuring the product variant master. In other words, it should follow the flow of materials through the cement factory. The advantage of following the flow of materials is that the resulting model is intuitively easier to follow for both F.L. Smidth's personnel and customers. Moreover, customers' requirements for the cement factory are often expressed in the context of material flow.

The idea of structuring the product variant master and thus the configuration system according to the flow of materials arose in connection with the construction of the product variant master. It was agreed to after a discussion of which customer requirements were important in relation to the configuration system, and which internal structure would be appropriate in the product variant master and the configuration system.

During construction of the product model, many useful discussions arose about how individual parts of the cement factory are constructed and connected with the other parts. In this way, the product variant master and the CRC cards have acted as a contract between the group responsible for making offers and the mechanical specialists with respect to which possibilities the offer group could provide for F.L. Smidth's customers. The development of the product variant master has also led to several improvements in various parts of the cement factory. An example, of this is the design of the cement mill' where the mill's design, for example with respect to gears and bearings, has been changed and optimized, which has made it possible to reduce the number of different cement mills from about 100 to 8.

During the development of the product variant master, a number of details concerning the individual nodes were described with the help of the CRC cards described in chapter 3. The product variant master was developed through a series of meetings between the configuration group and F.L. Smidth's mechanical specialists. Between meetings, the individual participants collected additional information and contributed to filling in the details on the CRC cards.

The product variant master contributes to producing an overall view of both the cement factory's and the configuration system's design. In addition, the product variant master is the starting point for the structure of the class model in the next phase, which involves a detailed analysis and construction of an object-oriented model (OOA).

Thus, CRC cards are used both in connection with the product variant master in phase 2 and the object-oriented class model in phase 3.

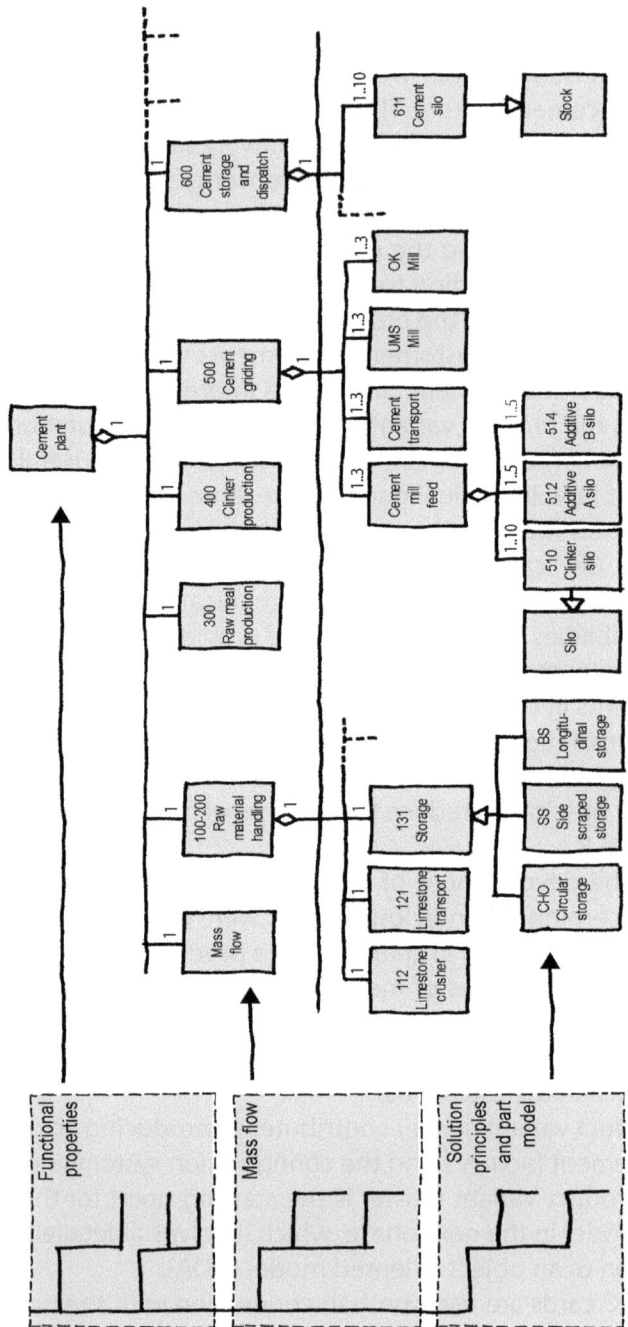

Figure 9.14. *The OOA model's two subject layers. Above: functions, properties and mass flow. Below: solutions in principle and part model.*

Phase 3: Object-oriented analysis

The next phase involves developing an object-oriented class model based on the structure in the product variant master. The individual nodes in the product variant master are modelled as object classes in the class diagram, if it is judged that the node can justify the creation of a separate object class. If there are no methods under the node in question, then the node is created as an attribute in the object class above. The same can be done if the node only contains very few, simple methods. In this case, the node is made part of the object class above, and both the attributes and the simple method or methods are created as attributes and methods in the object class above.

In connection with object-oriented analysis, the overall model is divided into two subject layers: a product subject layer, which contains the cement factory's machines and solution principles, and a subject layer containing the scope and mass flow for the cement factory (functions and properties), corresponding to the structure of the product variant master.

A section of the class model is shown in figure 9.14, which also shows how the two subject layers are related to the content of the product variant master. Figure 9.15 shows the two subject layers, together with the IT systems with which the configuration system is integrated.

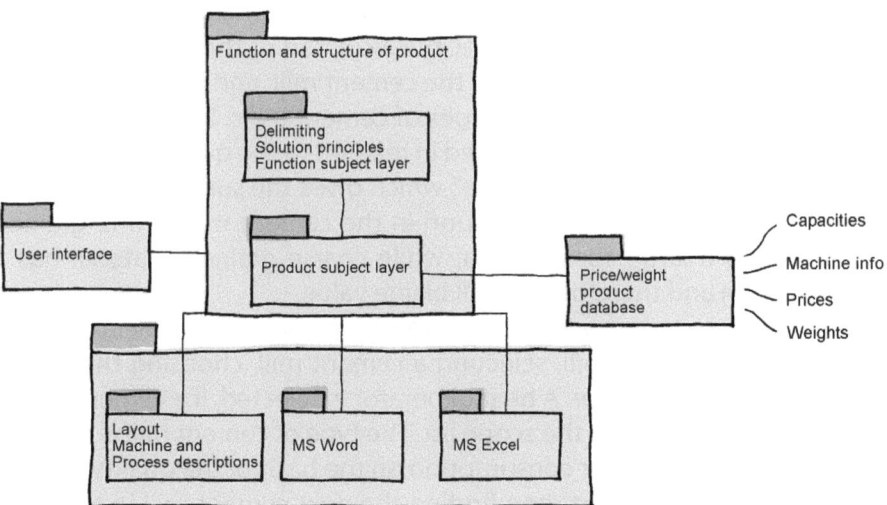

Figure 9.15. The model's subject layers and integration with other systems.

The configuration system is not only integrated with the user interface, but also with Microsoft Visio, in which a macro draws a process diagram. The configuration system is also integrated with Microsoft Word, which creates the actual written offer. Finally, the configuration system is integrated with a database containing layout drawings and machine and process descriptions, and a database with information about the individual machine's capacity, weight, price etc.

Figure 9.16 shows a section of the class diagram for the two subject layers for product and function. The class model includes both a number of part-of structures such as handling of raw materials, which contains the classes limestone-crusher, limestone-transport and storage, and also a number of kind-of structures such as stores, which are found in three different types, the circular store, sideways scraped store and longitudinal store respectively.

Figure 9.17 shows an example of a CRC card for the "OK Mill" class. On the CRC card, the class name, author/person responsible, date of origin, and latest revision are all specified, together with the relation to the rest of the model, i.e. the class is placed in an aggregation structure with the class "Cement grinding" as superclass. The class' mission is also specified by a short explanation in ordinary prose describing "what the class does". In this case, the class determines the machine type on the basis of the capacities calculated by the Mass Flow class.

The CRC card specifies under "attributes" some of the characteristics that describe a cement mill - in this case, the power consumption of the cement mill, the production capacity calculated in the Mass Flow class, the actual production capacity of the cement mill, and a table giving the production capacity of various types of cement mills. In addition to this, the power consumption is specified in relation to the quality of the materials, expressed in terms of "blaine", which gives the surface area per unit weight of the material to be ground in the cement mill. A fine-grained material has a large surface area, while coarse-grained material has a small surface and therefore a small blaine value.

Under product methods, examples are given of rules for calculating the power of the cement mill, selecting a cement mill, choosing the size of gears and heat generator. A heat generator is selected, if a cement grinding station is selected in the scope list. The type of cement mill is selected by calculating the power consumption on the basis of the quality (blaine) of the raw materials, and then finding the next cement mill in the table with a power capacity larger than the calculated power consumption.

The individual classes and the CRC cards' attributes are given well-

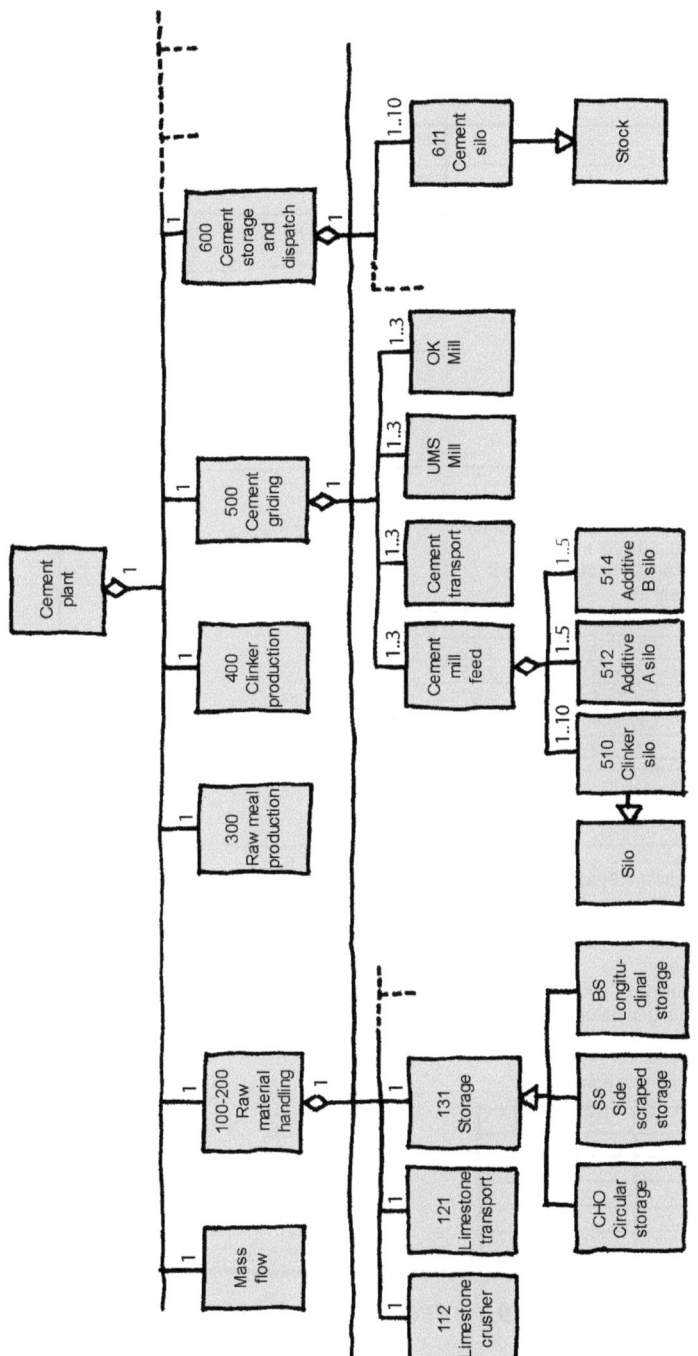

Figure 9.16. A section of the class diagram.

Class name: OK Mill	Date:	Author/version: LiHe

Class responsibilities: The class' responsibilities are to determine the machine type from the capacities calculated in mass flow.

Aggregation:	Generalisation:
Superparts: Cement grinding	Superclasses:
Subparts:	Subclasses:

Sketch:

Attributes:
Power Baine [0..1000] kW/ton
Production (mass flow) [0..10.000] ton
Power Mill [0..50.000] kW
Production Mill [0..10.000] ton/hour
Power Cement Mill

Type	Power [kW]
OK 27,4	1.636
OK 30,4	2.136
OK 33,4	2.970
OK 36,4	3.713
OK 39,4	4.554

Class collaborates with:

System methods:

Product methods:
Power Mill (calc) = Production Mass Flow * Power Blaine
Mill type is selected as the first mill in the Power Cement Mill table with a power large than the calculated power.

Heat generator
Heat generator only for cement grinding stations. Cement grinding stations in scope causes a heat generator to be selected by default.

Gear
Gear power >= mill power

Mass flow

Gear	Power [kW]
DMHG 18	2.690
DMHG 22	4.100
DMHG 25,4	6.500
2xDMHG 22	8.400
2xDMHG 25	10.000

Figure 9.17. *CRC card for the class "OK Mill."*

known domain names which are used in the daily work. This makes it easier for F.L. Smidth's employees to understand the model and to discuss and evaluate its contents.

The decision was made not to express methods in a formal notation or pseudocode. Instead, an effort has been made to express product methods etc. in ordinary language, using formulas and tables. This has been done in order to make sure that the model is easy to understand for mechanical specialists and other employees at F.L. Smidth.

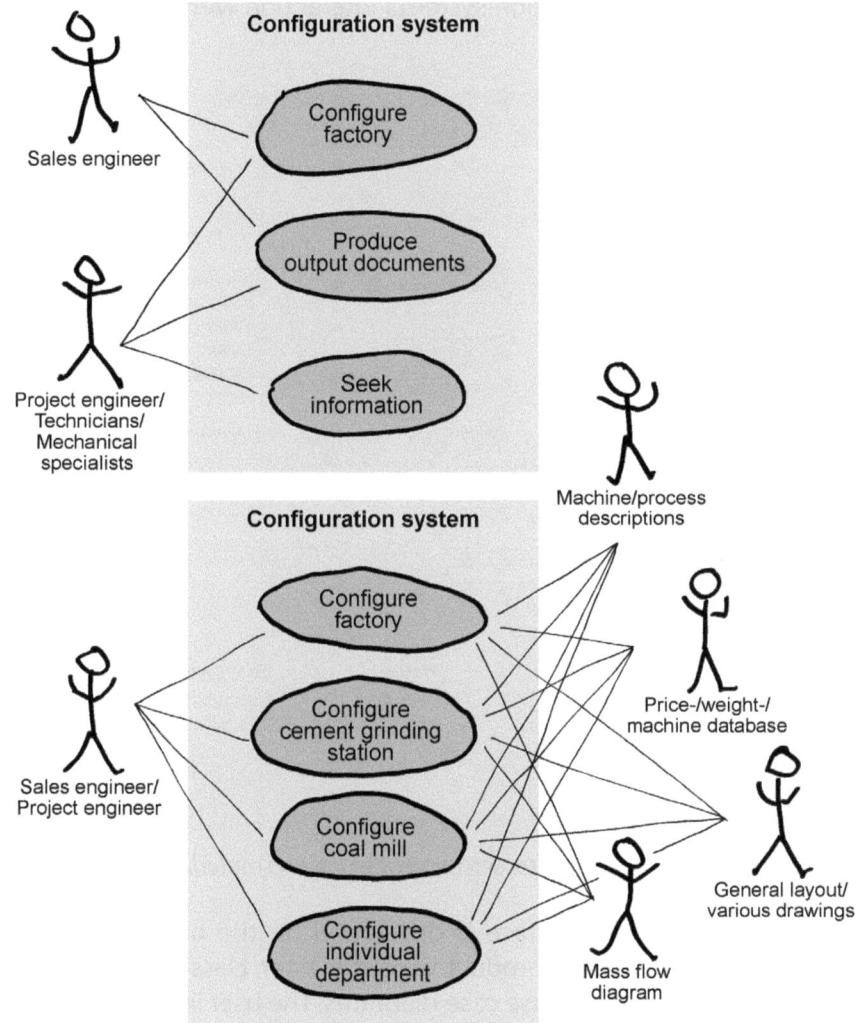

Figure 9.18. *Use case diagrams.*

To define the system's user interface, both early prototypes of the system's user interface and a series of use case diagrams were developed. Use case diagrams are primarily developed in order to define the individual user's requirements and expected use patterns when using the configuration system. Attention was also focussed on discovering the configuration system's interaction with other IT systems, such as the machine database or the system for drawing mass flow diagrams.

Figure 9.18 shows two examples of the system's high-level patterns of use, describing respectively the users' interaction with the configuration system and the configuration system's interaction with other, external IT systems.

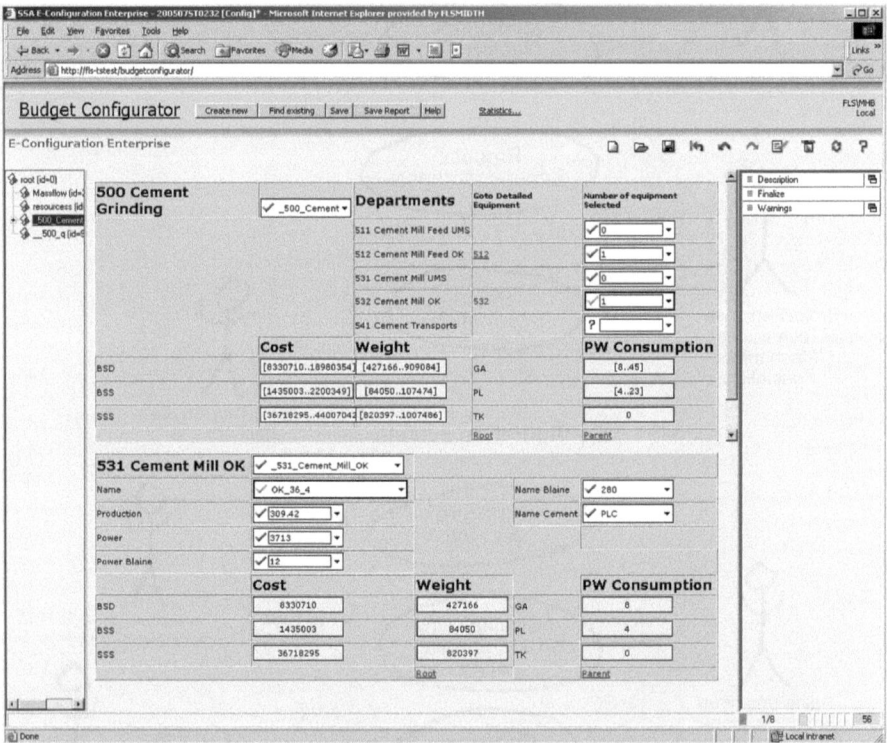

Figure 9.19. *Screen shot for selection of scope and material specifications.*

The system's user interface is developed on the basis of the model which has been created (product variant master, class diagram and CRC cards), together with the use case diagrams. The user interface is also designed according to the principle of following the flow of materials in the cement factory; thus, one of the first screen shots concerns the supply of materials to the factory. The next step is to specify requirements for the individual parts of the cement factory according to the cement factory's material flow; finally, the way the cement is transported from the cement factory is specified.

Screen shots also relate to the cement factory's overall conditions, such as the quality of the raw materials, operation times, emissions etc. figure 9.19 shows an example of a screen shot in the configuration system, where the input is cement grinding, equipment and requirements for the cement mill, and then on this basis, the configuration system selects the right mill. In addition, price and weight are given, together with the expected consumption of resources at F. L. Smidth, specified as number of man-weeks (PW).

Phase 4: Object-oriented design. Choice of software

In this phase, F.L. Smidth have amongst oher things selected standard software for implementation of the configuration system. The selection process took place as described in chapter 8. F.L. Smidth chose to use the product, E-Configuration Enterprise, from the company Infor.

The configuration system is object-oriented, which amongst other things means that the class model from Phase 3 can be implemented directly in the configuration system. This not only makes the programming task easier, but also means that the class model and the CRC cards can be used directly as documentation for the configuration system. Furthermore, the class structure means that it is possible to change a single class (particularly, classes placed at the bottom of the hierarchy) without this having any significance for other classes. This makes it easier to develop the configuration system further as time goes by.

Phase 5: Programming

The configuration system was programmed by an F.L. Smidth employee working full time on the programming task. The programming was done using the product variant master, the class diagram and the CRC cards as basis. The person programming the system was involved from the start of the project and developed a number of prototypes. He programmed and tested parts of the system when it was not certain how the relevant rules and calculations could be implemented.

Figure 9.20 shows the various windows used in connection with programming a configuration system in E-Configuration Enterprise.

Project Explorer includes an overall view of the configuration systems (projects) and each configuration system's modules (corresponding to themes in the class model) and the individual classes in the model. In Project Explorer, it is possible to add and remove classes. It also shows general properties, which hold for the entire model.

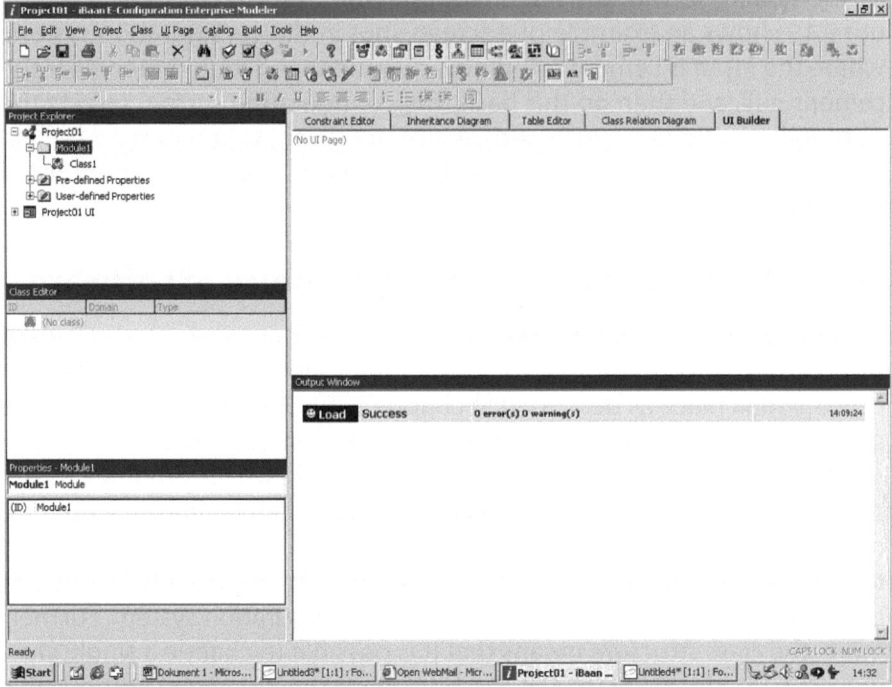

Figure 9.20. *Windows in the configuration system E-Configuration Enterprise.*

In Class Editor, it is possible to add and edit attributes of the individual classes. An attribute is defined by its type (boolean, identifier, integer, real or string), together with possible values.

Under Properties, it is possible to edit the properties of the individual classes, such as whether the class' attributes and methods are private or accessible for all the remaining classes, or to change the type of constraints.

In addition, a graphic module makes it possible to edit class structures within an Inheritance Diagram and a Class Relation Diagram, respectively. It is also possible to graphically edit constraints and work with tables and the system's user interface.

Figure 9.21 shows a section of the class structure in the configuration system.

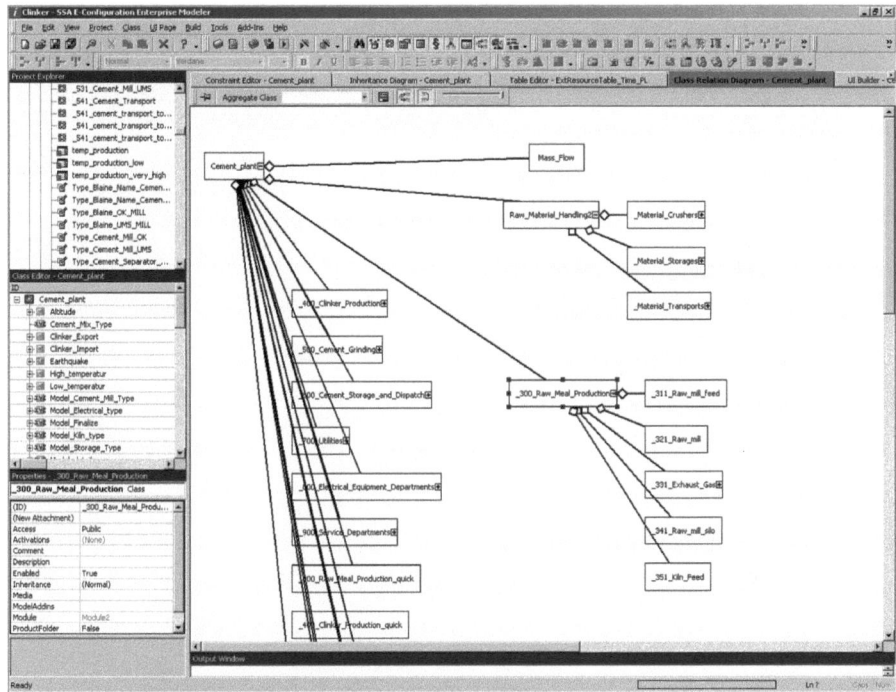

Figure 9.21. *A section of the configuration system's class structure.*

Figure 9.22 shows part of the configuration system's attributes and methods from class 531 Cement Mill OK. At the top left of the screen shot, the class' relationships with the other classes in the model can be seen. In the middle, the class' attributes (such as capacity or power consumption) are shown, and at the bottom the class' properties. On the right, examples of rules, such as a rule for calculating power consumption, is shown.

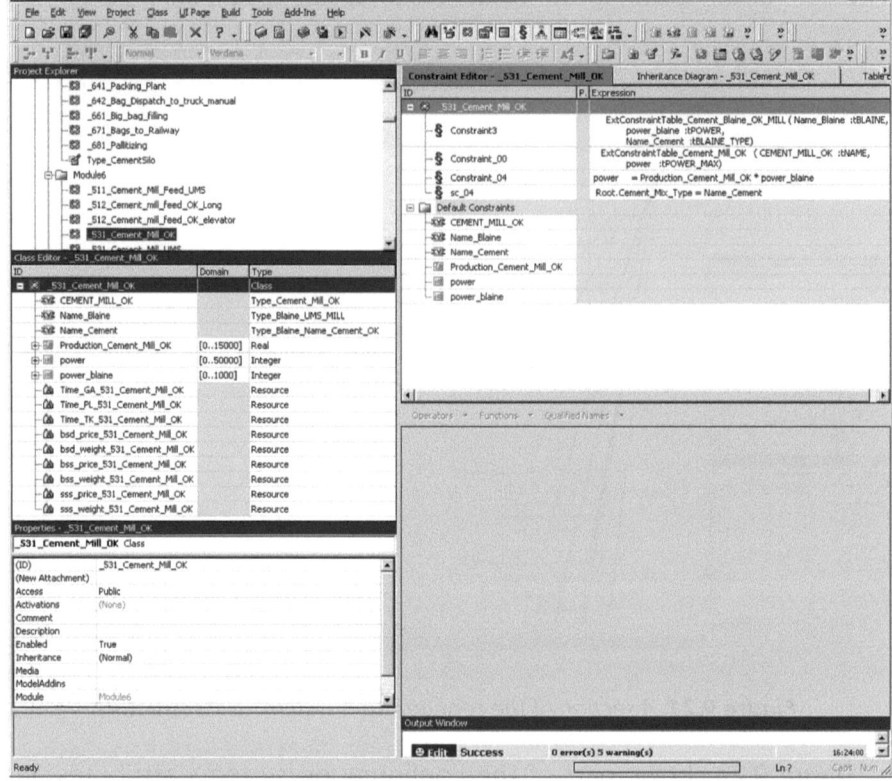

Figure 9.22. *Example of attributes and methods in the configuration system.*

Phase 6: Implementation

The first version of the configuration system was implemented within the organisation in 2000. Before the system was implemented, extensive testing of the system was performed, in order to ensure that the system was as free of errors as possible. In addition, a series of tests of the system with selected users were carried out, in order to receive feedback about the functioning of the system and the user interface, and in order to achieve gradual acceptance of the system. In connection with the implementation of the system, a number of courses were carried out for the users of the system on how the configuration system is organised (explaining the rules for dimensioning cement factories that are incorporated into the configuration system) and how users can use the configuration system.

In connection with the first version of the configuration system (scenario 1), there were also two super-users, who were responsible for developing the configuration system, and who therefore were in a position to advise the system's users on how to use the configuration system.

During implementation of the configuration system, it became apparent that the system can be used in several ways and for several purposes. Some of the most common ways of using the configuration system are as follows:

- A salesman can (possibly together with a super-user) use the configuration system to work out a budgetary offer for a customer.
- A salesman collects a group of specialists and a super-user. Together, during the course of 2-3 meetings lasting 1-2 hours, on the basis of input available from the customer, they determine the necessary inputs and configure a cement factory that they believe can best satisfy the customer's requirements. This way of operating ensures that the specialists' input and wishes are included, even in the first budgetary offer.
- Sales staff, specialists or others use the configuration system as a reference manual where they can find information, for example about solutions in principle, layouts, prices or capacity for various parts of the cement factory.

The new version of the configuration system (scenario 2) described in this chapter was implemented within the organisation in the course of Spring 2005.

Phase 7: Maintenance and further development

After the original project group was disbanded in 2000, the company set up a configuration team, which had responsibility for operation and further development of the configuration system, and an advisory group. The advisory group had the job of collecting and discussing experiences from using the configuration system, and of making contributions to continual updating of the system.

The configuration team consists of staff members who have detailed knowledge of F.L. Smidth's products and business processes, together with an employee whose competence lies in the areas of programming the configuration software.

The responsibility for updating the knowledge incorporated in the configuration system lies with the individual departments. The advisory group helps the configuration team to follow up on the mechanical specialists in the individual departments in order to keep the knowledge in the configuration system up to date. In the period 2000-2002, about 40-70 man-weeks per year were used to operate and maintain the configuration system. Some of these resources were used for collecting information and preparing scenario 2 for the system.

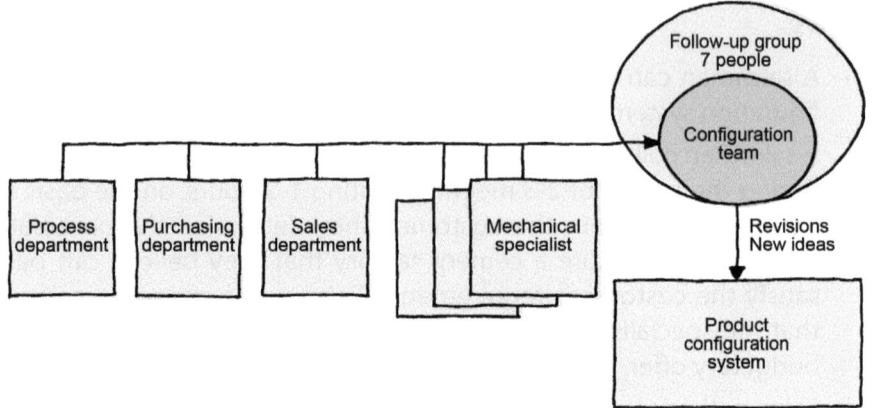

Figure 9.23. *Organisation of the maintenance phase.*

The product variant master and CRC cards are used to document the knowledge included in the configuration system. The class model was developed from the product variant master, but is used exclusively in connection with the programming of the configuration system. This means that the dialogue with the mechanical specialists and others takes place using the product variant master and CRC cards. Version control of the documentation and configuration system is handled by using version numbers.

Conclusion

The starting point for implementation of a configuration system for F.L. Smidth was recognition of the fact that the time used for making offers was increasing, while at the same time customers were asking for more offers. There was also a considerable demand from the customers for being able to receive offers within a shorter time.

The first step in the project was to carry out a Master's thesis project, in which a prototype of a configuration system was constructed. The prototype, which was finished in 1999, worked both as a proof of concept and as a means of developing the necessary competences at F.L. Smidth, so that a configuration project could be carried out.

Immediately before starting the configuration project, F.L. Smidth carryied out a large IT project, which did not achieve the expected results. This was one reason why, in specfying the task, management aimed at a small project with only 2-3 participants and a "stand alone" configuration system, which was not integrated with the rest of the company's IT systems. The slogan for the project was "keep it simple", and the project group had the important task of focussing on the 20% of the cement factory which is the basis for 80% of the costs and performance; therefore, details that did not critically affect the cement factory's price and performance were not included in the configuration system.

The configuration system was taken into service in 2000, after extensive tests and debugging that ensured that the configuration system was close to being error-free. The system was also tested with selected users, and the sales personnel who were to use the system were trained.

Experience with use of the configuration system has been positive. Initially, the sales personnel were hesitant about using the system. This problem was solved by having a member of the configuration team work together with the salesman, so they could perform the configuration together. Figure 9.24 shows the volume of budgetary offers made in 2003 for which the configuration system was used.

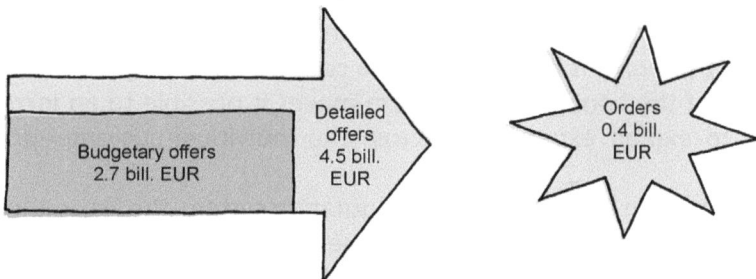

Figure 9.24. *Volume of budgetary offers for which the configuration system was used in 2003.*

Budgetary offers for about 2.7 billion euro and detailed offers for about 4.5 billion euro were produced. The result of this was orders worth about 0.4 billion euro. The size of the orders was typically from 25-40 million euro to 140 million euro. Total costs of the configuration project until the year 2000 were about 800,000 euro. Operation of the configuration system costs about 1-2 man-years per year. Use of the configuration system has meant that F.L. Smidth has been able to cut the use of resources for making offers in half, while the time interval from customer enquiry until the final contract is signed has been markedly reduced. Use of the configuration system has also meant that F.L. Smidth is able to respond to all enquiries with an offer. A single extra order to F.L. Smidth can cover the total project costs several times over.

Use of the configuration system has also led to a number of improvements in the sales process:

- Negotiations with the customer have a better structure.
- The configuration system ensures that the salesman collects all the necessary information before a budgetary offer is sent.
- The use of default values means that it is possible to make an early offer using very little input from the customer.
- Customer enquiries are answered more rapidly.
- Almost all enquiries are answered with an offer.
- It becomes possible to simulate different solutions for the customer.
- The configuration system can optimize the cement factory with respect to selection of previously produced parts and selection of parts produced by FLS companies.

Using the configuration system also means that sales personnel avoid burdening the specialists in connection with making budgetary offers. The most important benefit from the configuration project, however, is that use of the configuration system makes it possible to an increasing extent to exploit experiences across the individual customer-oriented projects.

In connection with using the configuration system, the focus is increasingly on defining and exploiting modules that can be used in the context of dimensioning the individual cement factories. In other words, the cement factory is dimensioned by selecting and adapting the modules that were developed to fulfil a spectrum of customer requirements. The modules that were developed thus become pre-requisites for developing the configuration system itself. At the same time, the configuration system ensures that the modules developed are used in the individual customer orders.

Figure 9.25 summarizes the most important costs and benefits from using the configuration system.

Costs for development of the configuration system include both the time used for development and specification of modules, creating the product variant master, CRC cards and the class model, and programming the product configuration system; and the cost of the software. Costs for the operation and maintenance of the configuration system include the time spent on making continual updates of the product variant master, CRC cards and class diagram, changes in the programme, and software licenses.

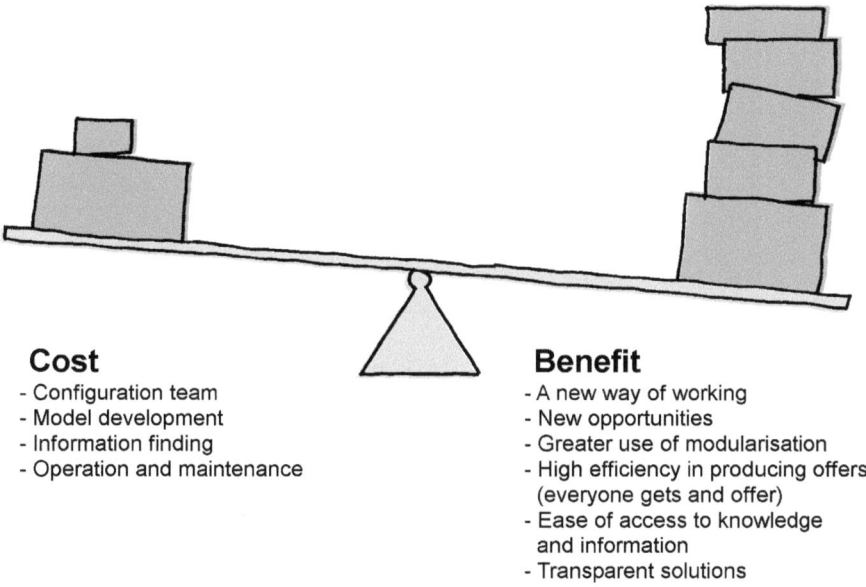

Cost
- Configuration team
- Model development
- Information finding
- Operation and maintenance

Benefit
- A new way of working
- New opportunities
- Greater use of modularisation
- High efficiency in producing offers (everyone gets and offer)
- Ease of access to knowledge and information
- Transparent solutions

Figure 9.25. *Costs and benefits from using the configuration system.*

The configuration system has led to a new way of working at F.L. Smidth. The staff has achieved increased awareness of the possibilities for exploiting experiences which cut across the jobs for individual customers. To strengthen this benefit, a department for modularization and configuration has been established.

Authors

Lars Hvam

Lars Hvam, Ph.D., is an Associate Professor at the Technical University of Denmark. He has been working on product configuration for more than 15 years as a teacher, a researcher and as consultant for more than 15 configuration projects in large industrial companies. He has supervised eight Ph.D. projects on the construction and application of configuration systems and has been the project leader for four large research projects on product configuration.

Lars Hvam is also the founder and current chairman of the Product Modelling Association, whose aim is to disseminate knowledge of the possibilities offered by product configuration through courses and other activities.

Niels Henrik Mortensen

Niels Henrik Mortensen, Ph.D., is an Associate Professor at the Technical University of Denmark (DTU). He has been engaged in research into and teaching of product configuration for 10 years.

Niels Henrik Mortensen has also been a consultant for more than 15 configuration projects for companies in Denmark and abroad. At DTU, he is the supervisor for 6 Ph.D. students within this field.

Niels Henrik Mortensen is a member of the board of the Product Modelling Association.

Jesper Riis

Jesper Riis, M.Sc., Ph.D., HD(O) has worked with product configuration for more than 8 years, both as a researcher and as consultant in a long series of companies. He is currently employed at the engineering company GEA Niro A/S as group manager of the group "Engineering Business Solutions".

Index

MIX
Papier aus verantwortungsvollen Quellen
Paper from responsible sources
FSC® C105338

If you have any concerns about our products,
you can contact us on
ProductSafety@springernature.com

In case Publisher is established outside the EU,
the EU authorized representative is:
**Springer Nature Customer Service Center GmbH
Europaplatz 3, 69115 Heidelberg, Germany**

Printed by Libri Plureos GmbH
in Hamburg, Germany